# BIOTECHNOLOGY IN FUTURE SOCIETY

EUROPEAN FOUNDATION FOR THE IMPROVEMENT OF
LIVING AND WORKING CONDITIONS

# BIOTECHNOLOGY IN FUTURE SOCIETY

## Scenarios and Options for Europe

Edited by
EDWARD YOXEN
Department of Science and Technology Policy
University of Manchester

VITTORIO DI MARTINO
European Foundation for the Improvement
of Living and Working Conditions
Dublin

OFFICE FOR OFFICIAL PUBLICATIONS
OF THE EUROPEAN COMMUNITIES
Luxembourg

DARTMOUTH
Aldershot

Publication No EF/88/08/EN of the European Foundation for the Improvement of Living and Working Conditions, Loughlinstown House, Shankill, Co. Dublin, Ireland.

British Library Cataloguing in Publication Data
Biotechnology in future society: scenarios and options for Europe.
1. Biotechnology. Social aspects.
I. Yoxen, Edward. II. Di Martino, Vittorio
303.4'83



Office for Official Publications of the European Communities, 2 rue Mercier, L-2985 Luxembourg.

ISBN 92-825-8597-2 Catalogue Number SY-52-88-598-EN-C

Dartmouth Publishing Company Limited,
Gower House, Croft Road, Aldershot,
Hants. GU11 3HR, England.

Gower Publishing Company,
Old Post Road, Brookfield, Vermont 05036,
USA.

ISBN 1-85521-016-9

Typeset by Byte-Bac (Manchester) Ltd. and
Printed and bound in Great Britain by Billing and Sons Limited, Worcester.

CONTENTS

# Preface

by Phili Viehoff, MEP

It is an honour to be invited to speak at this international seminar organised by the European Foundation for the Improvement of Living and Working Conditions. The fact that I am here for the second time within a year shows the strengthening of the links between the European Parliament and the European Foundation. I am very pleased about this development and I hope that it will be continued, because I think it is in the interests of both organisations.

As Rapporteur for biotechnology in the European Parliament since 1984 I have benefitted from the international research that has been organised by the European Foundation. Despite the importance of this research, there are still many important questions that remain unanswered over the wide field of biotechnology. Moreover many more questions are emerging as the diffusion of biotechnological products and processes begins.

One of the conclusions which emerged from the research on biotechnology funded by the European Foundation is that at many levels our interaction with the technology is very much a learning experience. We need to learn much more about the complexity of the new biotechnologies and its possible medium and long term impacts. Only by learning about these trends can we develop adequate policies to support biotechnology in socially responsible and desirable ways.

There is therefore an urgent need for future-oriented, forecasting and assessment studies of the social and economic dimensions of biotechnology. I strongly recommend that this kind of research should get utmost priority on the research agenda of the European Foundation.

Let me now say something briefly about the activities of the European Parliament in the field of biotechnology. In response to proposals from the European Commission to set up biotechnology programmes, the European Parliament has discussed in several reports, working documents and in hearing with experts, the potential, the utility, the limitations and the possible risks of biotechnology.

When we started to discuss biotechnology for the first time in 1980 not many politicians were really interested in the subject. This was not only the case at the European level, but also at the national level. Possibly the Federal Republic of Germany and somewhat later Denmark were exceptions to this rule. But in general biotechnology had to fight for its place in political debate, although in scientific and administrative circles planning for the new biotechnology was going forward at high speed.

But along the way biotechnology got more and more attention in the European Parliament, not least because in the reports we prepared for the Committee on Energy, Research and Technology, we took a very broad view. In our reports we showed the possible impact of biotechnology in all kinds of policy areas, from agriculture to health care, from environmental regulation to the problems facing less developed countries.

We also stimulated other Parliamentary committees to think about the particular impact biotechnology could have for their own areas of concern. So in 1985 we organized a hearing to bring together the different Parliamentary committees and to discuss the impact of biotechnology with experts from all over Europe. I am pleased to say that some of the experts who took part in that hearing are here today as well. In 1986 biotechnology eventually became a subject for seven committees in the Parliament and now several, more specialised reports have been drafted, for instance by the Legal Affairs Committee and the Committee on Agriculture. In January 1987 the main report, *Biotechnology in Europe: The need for an integrated policy* was discussed in the plenary session in Strasbourg and in February 1987 the Parliament accepted the report and its (amended) resolution.

Within my own political group we have set up a biotechnology working group to exchange information from the different committees. In a recent publication of our group, a brochure called, *Biotechnology: A challenge full of promise and pitfalls*, we recommend the establishment of an interparliamentary working party on biotechnology. Such an interdisciplinary working party could be of great help in planning for a more coherent policy for biotechnology at the European level. For instance, it does not make much sense to study ecological problems without knowing the European judicial framework, within which solutions must be sought. The same can be said for the social and economic dimensions of biotechnology. We have to learn how the different aspects relate to each other; which aspects are of most importance and why; what the consequences will be of the unequal development of biotechnology in the different Member States and how we can prevent European disintegration. The last aspect relates particularly to the possible effects that biotechnology may have on European agriculture, a subject already reviewed by the European Foundation.

The difficulty we had in the Parliament in writing our reports, and no doubt many of the experts here have experienced the same problems, is that with a complex set of technologies it is difficult to foresee all the effects and consequences at an early stage. The concrete facts do not exist yet or are still secretly held in laboratories. So the construction of reliable scenarios for the future depends entirely on close reading of industrial and scientific material and the disciplined exercise of the imagination. If that is not done, and if any effort at forecasting is delayed until the technology has reached an advanced stage of development, then it will probably be too late to intervene effectively to influence subsequent developments.

In daily political life we have to balance the need to encourage promising new areas of research and technological application on the one hand against the need to enforce every reasonable safety precaution both in the workplace and to protect the environment if genetically engineered organisms are released into it. The emphasis here on balance is also central to an enticing new development in which biotechnology would play a crucial role. From our research and that carried out by others, it is easy to show that the increased concentration of power over the combined agriculture and food-supply chain, together with a trend to larger and larger units in the dairy and stock-rearing sectors, is placing political and economic relationships both within the Community and outside under increasing strain. Similarly in health care organising the production of services in larger and larger units has led not merely to 'white elephants' but even to 'white mammoths'. These unworkable health factories, which often themselves create disease, consume vast amounts of resources and often act against the very goals they were set up to advance. Such trends then display a continuing move away from balanced systems of food production and health care.

By contrast with this one can consider the prospect for decentralisation. The use of new information technologies linked to those of biotechnology could replace the old obsession with economies of scale by new principles of flexibility. For example biotechnology could make regions less dependent on agricultural imports, because the diversity of cultivation could be increased and the processing of crops be done much more intensively than is now the case. An important consequence of reduced dependency would be more political freedom of manoeuvre at the regional and local levels. It is a political challenge to use that leeway to give specific content to decentralised decision-making. Several of the new European technology programmes are aimed at backward regions. With the installation of new communications networks these regions could very well make more use of biotechnology and become more integrated into the Community, without giving up their historic characteristics. Also in industry it would be possible to apply biotechnology in such a way that, instead of increased centralisation and hierarchy of control, more scope is provided for self-determination in the workplace.

The terms in which biotechnology eventually becomes established ultimately depend on the balance of political and economic power. However, and I must stress this, the manner in which biotechnology is consolidated can certainly be influenced and is not preordained.

Before we can enter a stage of decentralised development in biotechnology, much research has to be carried out. This research is also important to stimulate public debate. For this reason better dissemination of research is essential. This is true also of research carried out by the European Foundation. Although scientists and researchers very often prefer to write for established journals, they should be aware that these only reach a very limited number of people, who are very often already well informed. Therefore it is necessary to translate the results of research into accessible language and into an appealing form to increase awareness among a wider public. In this meeting many experts discussed the latest information about the social implications of developments in biotechnology. For it is 'learning by learning' that can help us increase the social benefits of biotechnology for society at large in a responsible way. Unfortunately we cannot afford to stay in the 'learning by learning' phase. Strong industrial and research lobbies in the Community and outside are pressing for 'learning by doing' and 'learning through application'. If progress towards greater use of biotechnology is not to be frustrated, when all possible safety precautions have been considered, this pressure for continued innovation has to be recognised and further support for biotechnology made possible. But this must be only on one condition: that parallel to the application-oriented activities more research on the social impacts in the broadest sense must be carried out. In other words, I make a plea for an intensification of social and economic research to assess the impact of biotechnology on living and working conditions. As I said earlier, the European Foundation can play an important role in these assessment studies. I hope that this seminar will strengthen the view that research that builds on the base of expertise already established by the Foundation in this area should be carried out in the years to come.

# List of European Commission, Parliamentary and European Foundation reports on biotechnology

List of reports on biotechnology from the European Commission, the European Parliament and the European Foundation for the Improvement of Living and Working Conditions.

European Commission

Proposal for a Council Decision adopting a multiannual research action programme of the EEC in the field of biotechnology (1985-9). COM (84)230

Proposals for Council directives on the approximation of national measures relating to the placing on the market of high technology medicinal products, particularly those derived from biotechnology. COM (84)437

Stimulating agro-industrial development: discussion paper. COM (86) 221

Review of the Multiannual Research Programme COM (86) 272

A Community Framework of the Regulation of Biotechnology COM (86) 573

Biotechnology Action Programme - Progress Report 1987 Vol. 1 An Overview CD-NA-11-138-EN-C; Vol. 2 Scientific Papers CD-NB-11-138-EN-C.

Proposal for a Council Decision to adopt a first multiannual programme (1988-93) for biotechnology-based agro-industrial research and technological development - ECLAIR. COM (87) 667

European Parliament

Synopsis of the Hearing on Biotechnology: Committee on Energy, Research and Technology (Brussels: CEC, 14 January 1986).

Report on behalf of the Committee on Energy, Research and Technology, Biotechnology in Europe and the need for an integrated policy. Rapporteur: P. Viehoff. Document A 2-134/86.

Report on behalf of the Committee on Agriculture, Fisheries and Food, on the Effects of the use of biotechnology on the European farming industry. Rapporteur: F.W. Graefe zu Baringdorf. Document A 2-159/86.

A. Roobeek, Biotechnology: A challenge full of promise and pitfalls, (Brussels: Socialist Group, European Parliament, 1987)

European Foundation.

The impact of biotechnology on the environment ISBN 92-825-7529-2

The impact of biotechnology on working conditions ISBN 92-825-7528-4 (in Italian)

The international dimension of biotechnology in agriculture ISBN 92-825-7531-4 (in Dutch)

The social impact of biotechnology ISBN 92-825-7530-6

The impact of biotechnology on living and working conditions - Consolidated report
ISBN 92-825-7532-2

# List of Contributors

Dr. David Banta,
Professor of Health Economics
Rijksuniversiteit Limburg,
Maastricht
The Netherlands

Pascal Bye,
INRA-IREP,
BP47X,
38040 Grenoble Cedex,
France.

Dr. Peter Daly,
Biotechnology Group,
National Board for Science and
Technology,
Shelbourne House,
Shelbourne Road,
Dublin 4
Eire.

Dr. Nadine Fresco,
U158 INSERM,
Hopital des Enfants Malades,
149 rue de Sevres,
75015 Paris,
France.

Prof. Giuseppe Lanzavecchia,
ENEA,
Viale Regina Margherita 125,
0098 Rome,
Italy.

Richard Pearson,
Institute of Manpower Studies,
University of Sussex,
Falmer, Brighton,
United Kingdom.

Jorgen Lindgaard Pedersen,
Institute of Social Sciences,
Technical University of
Denmark,
Building 301,
DK-2800 Lyngby,
Denmark.

Wolf-Michael Catenhusen,
Bundestag,
Bonn, Federal Republic Germany

Prof. Gerd Junne,
Universiteit van Amsterdam,
Vakgroep Internationale,
Betrekkingen Volkenrecht,
Herengracht 510,
1017 Amsterdam,
The Netherlands.

Dr. Edward Yoxen,
Department of Science and
Technology Policy,
University of Manchester,
Oxford Road,
Manchester. M13 9PL,
United Kingdom.

Vittorio Di Martino,
European Foundation for the
Improvement of Living and
Working Conditions,
Loughlinstown House,
Shankill,
Co. Dublin,
Eire.

Jos Bijman,
Universiteit van Amsterdam,
Vakgroep Internationale
Betrekkingen Volkenrecht,
Herengracht 510,
1017 Amsterdam,
The Netherlands.

Danielle Mazzonis,
ENEA,
Viale Regina Margherita 125,
0098 Rome,
Italy.

Anna Luise,
ENEA,
Viale Regina Margherita 125,
0098 Rome,
Italy.

# PART ONE: INTRODUCTION

# 1 The Context of this Volume

Edward Yoxen, Vittorio Di Martino

## THE BEGINNING, NOT THE END

In the last decade biotechnology has not gone unremarked. Indeed many of its more striking technical achievements, commercial events, and policy decisions have attracted a great deal of analytic comment. Such more or less detailed reviews have been informed by different political and economic assumptions and as a result their conclusions have been correspondingly varied. Taken together this material amounts to a substantial secondary literature, comprising reports by government committees and agencies, surveys by financial institutions, and critical commentaries prepared by representatives of particular lobbies or pressure groups. Why then add yet another volume to this material, if there is already an abundance ?

There are at least three cogent responses to such scepticism. The first is the actual volume of commentary is misleading. Much of the material is repetitive, some of it very poor in quality, and all of it is dating rapidly, as both the technologies and the politico-economic context are constantly being transformed. The second is that relatively little material has looked at developments from the perspective of the European Economic Community, taking into account the policies being developed and pursued at the European level. The third is that such commentary has usually been made in response to specific developments - a particular legal decision, a new business venture or a technical breakthrough - and has often been somewhat speculative. What is clearly needed now to complement the highly fragmented literature of reportage and rapid evaluation is research, intended to develop more general models and scenarios, and to test them. Biotechnology is now sufficiently well established as a technological and industrial phenomenon to support a critical debate between researchers about the deeper economic and political significance of readily identifiable technical trends and their putative effects. There is now enough material to support contending interpretations. Analysts of biotechnology need in their turn to expose some of their cherished speculations to more extensive evaluation.

This in itself is unusual and distinctive: these proceedings record a debate over the kinds of question that should be asked about the biotechnological future, as it is being formed. In fact there are two matters of significance here, one is the attempt to consider the future, obviously an exercise beset with problems created by the arrow of time; the other is the fact that debate took place between representatives of different interest groups over how the exercise should usefully proceed. The meeting then was a debate over future 'futures' research.

As one would expect real differences of approach, priority, perspective and political ideology emerged in the plenary sessions and in the working groups with more specific objectives. These are explored in more detail in the final essay in this volume. But despite these differences, there was also agreement that more prospective research on the social and economic impacts of biotechnology would be very valuable. The point then to emphasise here is that the papers which follow and the discussions they engendered in Dublin in November 1987 represent the formative phase of a research programme, not its conclusion. They clarify and underline what needs to be done.

## THE VIRTUE OF INDEPENDENCE

Anyone looking for data on the operation of the European Economic Community confronts the problem of apparent over-supply. There are vast mountains of material available in libraries all over Europe on many different aspects of economic and social life. If research was just a matter of extrapolating time series, then the task would be straightforward in principle, if laborious in practice. But the problems of technological forecasting and scenario construction are more subtle. They require the articulation of models of likely events, informed by some kind of prior theory, which indicates why particular phenomena are connected, why some will remain connected and why others will develop independently.

Exercises of this kind have been conducted by the European Commission. In particular the FAST programmes within DG XII have applied procedures for technological forecasting to the identification of future research needs. The committees of the European Parliament also periodically review specific areas of concern. The exercise reported here however was organised by an organisation independent of both the Commission and the Parliament, the European Foundation for the Improvement of Living and Working Conditions based in Dublin. The Foundation funds research on a wide range of issues that relate to employment, the quality of work, and the general social environment within the European Community. Its funding forms part of the Community budget, but as an organisation it enjoys complete formal autonomy to sponsor research as its administrators and advisory board see fit.

At the same time if such research is to be of relevance to continuing policy debates in Europe it must take account of Community policy and the political processes through which these policies are developed, contested and modified. This obviously requires the interest and cooperation of officials of the Commission and members of the Parliament. The seminar on biotechnology in Dublin was a striking instance of this. Mark Cantley of the CUBE Unit in DG XII, who has been closely involved with policy for biotechnology at the European level for a number of years, acted as chair for much of the meeting, and several parliamentarians, including Phili Viehoff, whose speech prefaces this volume, made extended contributions to the discussions. Cantley's skilled direction of the debate provided a constant flow of background information on how the various parts of the Commission interact, whilst the interventions of the MEPs helped to mitigate some of the academicism and formality of an international research workshop. Both forms of input helped to remind the participants that many of the issues under discussion were subject of controversy and sharp conflict in other settings.

By bringing together representatives of many different organisations and interest groups the Foundation was pursuing a tradition of fostering debate between the social partners to identify areas where research could moderate conflict and provide opportunities for learning and increased participation in decision-making in future.

## BIOTECHNOLOGY IN THE HISTORY OF TECHNOLOGIES

Biotechnology is a neologism of the mid 1970s. It connotes modernity, technological intensity, a basis in advanced research and commercialisation by sophisticated, future-oriented business enterprises. Yet its origins are ancient, extending, if one insists on tracing the lineage as far as possible, to the household procedures of food preservation, fermentation and crop selection innovated in agricultural communities up to six thousand years ago. At the very least this reference to the remoteness of the historical origins of biotechnology makes clear how relatively recent is its industrialisation, and the continuing importance of pre-industrial expertise and patterns of consumption. Several participants at the seminar argued that traditional agriculture operates with standards of ecological rationality, biological diversity, and consumer satisfaction which more modern, energy- and resource-intensive production does not meet. On this view only those biotechnologies rooted in traditional expertise and connected with non-intensive, environmentally benign production should even be countenanced.

Yet the more common view is that the continuing application of novel research ideas to industrial production, the essence of modern biotechnology, offers the prospect of real technical, economic, industrial and social progress, which it would be irresponsible to reject. In the late 1970s one version of this technological optimism was promoted with tremendous vigour by scientists, business managers and investors, with a personal stake in the new technology. Their sincerity is not in question here. It is enough to note that even in this quarter there is now more scepticism about the immediate commercial promise of biotechnology, given the considerable problems of innovating, financing and marketing new products and processes in the economic and regulatory context of the late 1980s. Even with the enormous technical boost offered by recombinant genetics (genetic engineering), by parallel developments in immunology and cell biology (monoclonal antibodies) and by information technology for data storage, data analysis, and process control, it is a very expensive and uncertain business to bring biotechnological inventions to the market. It can be done, as a range of new products demonstrate, but they are far fewer in number than many expected in the late 1970s.

It is now around fifteen years since the 'invention' of recombinant DNA research and hybridomas, which stimulated such interest in biotechnology, and caused a diverse set of industrial sciences to be re-named. It is ten years since the appearance of research-based companies principally in the United States specialising in the commercialisation of the new biotechnology, and five years or so since the beginning of serious investment by European governments in pre-competitive research. However the trajectories of technological development are still far from being fully formed, and debate over how biotechnology should be promoted is intensifying.

In particular in Europe its future will be intimately linked, in several different ways, with that of agriculture, as two papers in this volume make abundantly clear. Thus several interest groups at the moment are attempting to change the whole nature, distribution, intensity and purpose of agricultural production within the European Community. In this they are assisted by the enormous costs of agricultural price support, which have developed for historic reasons, but which now generate tremendous economic pressures for changes of policy. The promise of new crops, new processing technologies, and new agricultural products, which would result in a dramatically restructured agricultural base in Europe, also makes the possibility of change more plausible. But there are important economic, political and technical obstacles in the way. In particular some farmers feel they have no option but to resist such changes as vigorously as they know how, since they do not have the resources to survive in the changing markets for agricultural products. For them biotechnology connotes extinction.

Thus as applied research moves beyond the phase of inevitably simplistic rhetoric, intended to secure initial investment, into a phase of longer term planning, development and industrial consolidation, it is clear from a range of policy studies that major political conflicts have to be resolved, if innovation is to continue, in whatever direction. In some ways it was always obvious that this would be so. The confident rhetoric of the entrepreneurs of the late 1970s simply ignored such possibilities. We now stand at a point where more reflection, more information about possible paths of development and more critical examination of earlier speculations is highly desirable. The very fact that the technologies that constitute biotechnology are not set in a final form represents an opportunity for detailed research into its possible effects, to modify the plans now being laid for decades of innovation extending into the next century.

## SELECTIVITY ACKNOWLEDGED

The truly remarkable characteristic of biotechnology is the number of areas of economic and social life, on which sooner or later it will impinge. These include the much discussed areas of health care, the chemical industry and agriculture, as well as the less often mentioned areas of waste disposal and energy production. Within each of these very broad areas there are literally hundreds of different sub-areas, to which some form of biotechnology may make a significant contribution. Although some studies, such as that conducted by the Office of Technology Assessment of the US Congress in 1982, have tried to sketch in as much detail as possible, the only rational strategy with very limited resources is to be highly selective and to concentrate on the most immediate and concrete questions.

The exploratory research funded and organised by the European Foundation in 1985 and 1986 was already deliberately selective. After prior discussion with a number of experts, attention was focussed on four areas; medicine, agriculture, the environment and the future of work and employment. Each study was undertaken by a separate team over a six month period. The work on medicine was undertaken at the University of Manchester in the UK: that on agriculture at the University of Amsterdam: the environmental study was undertaken at the Technical University of Copenhagen and the work on employment at ENEA in Rome.

The resulting studies were published by the Foundation in 1987, along with a consolidated report which brings together the main conclusions into a single document. In addition bibliographical assistance was provided by the European Biotechnology Information Project at the British Library in London, and a separate cumulated bibliography on the social impact of biotechnology has also been published by the Foundation.

A number of general conclusions emerged from or were confirmed by this work. One was that the form of biotechnology in Europe is still very much dependent on political decisions yet to be taken. Another was that its development is inextricably linked with very significant structural changes in the European industrial base, in the health care system and in the agricultural sector. This question is considered again below. Another conclusion was that some developments were of immediate significance, others would only become important a decade or two hence. Thus biotechnology seemed likely to facilitate major changes in medicine, agriculture and environmental management, even in the short term, whilst the effects on bulk chemical production, on energy supply and on mature manufacturing industries appeared to be rather more distant, indeed almost over a time horizon that define the limits of useful speculation. Thus timescales alone offer a rationale for selective attention.

Finally it became apparent that the term 'social impact' covers a whole range of different phenomena and processes. Some of these are very concrete, such as the loss of markets or the displacement of jobs; others are much less tangible, but still real and important, such as changes in attitude, changes in legal or cultural convention and changes of approach to particular problems. At the same time the more concrete questions may not be accessible empirically, since the relevant data simply does not exist. This seems to be true, for example, of changes in the nature of work and in levels of employment. There is a great deal of intelligent speculation about the skills required for biotechnological production and whether those skills can be supplied from the existing labour force, but there is little hard information with which to decide between scenarios.

At the beginning of 1987, when plans were drawn up for the Dublin seminar to take such issues before a wider audience of policy-makers, politicians and academics, the question of selectivity arose again. To organise an international meeting on a wide range of issues would be unwieldy and inefficient, particularly if the objective was primarily to debate and refine a research agenda. The plan from the outset was to run the meeting as a working seminar, with specialised workshops in which invited experts would consider together possible future research in very specific areas. Obviously the selection of issues for the exploratory research could have been repeated in the Dublin meeting. To some extent this was the case.

Since the Foundation has devoted much attention to the issue of participation in discussions about the origination, design, commissioning and use of new technologies, and since several aspects of biotechnology have been the subject of intense controversy in recent years, it seemed important to treat participation in political debate and industrial negotiations as an issue in its own right. In the present context it was certain that any workshop on this topic would inevitably deal with debates on environmental protection. Indeed that see med an effective way of ensuring that discussions in Dublin focussed not on the technicalities of environmental risk and its possible control, but on the more general issues of what count as an acceptable risk and whose views on this topic carry weight.

Thus four workshops were placed at the core of the meeting: one on different areas of medicine, one on European agriculture, one on work, employment, training and workplace safety, and one on participation. Clearly this meant that some applications of biotechnology were not considered, notably the production of energy and the management of waste material. This is not to deny their importance. It simply reflects the priorities of an organisation with limited resources. No doubt for some people these are the most promising and interesting areas of biotechnology. If that is the case, one feels sure that they will easily find other venues in which their views can be advanced. But what is important here is simply to acknowledge that these proceedings are somewhat selective in their coverage of biotechnology, to indicate the selection was not arbitrary, although we can offer no ultimate or incontestable justification for it. Perhaps in fifteen years the decision not to pursue the energy questions in 1987 will seem incomprehensible: and perhaps not. Whatever happens the decision to pursue those relating to biotechnology in medicine and agriculture will surely have been right.

It is also important not to view the papers assembled here as definitive studies, to be judged by the standards applicable to a major technical review. They are too short for that, and were commissioned as concise, well informed, up to date surveys of particular areas, that could serve as the basis for workshop discussions. In this respect they functionned extremely well and provided a wealth of possibilities for further research. The resulting suggestions are summarised in the final paper in this volume.

# 2 Biotechnology in Europe: The Role of the Commission of the European Communities

Mark Cantley

These remarks are made in a personal capacity, and do not represent a statement of Commission policies.

## FROM A RESEARCH PROGRAMME AND A FUTURES GROUP TO A EUROPEAN STRATEGY

The Community institutions play a significant role in European biotechnology, and one strongly related to the many issues implied by the broad title of this forum - issues of research and technology, of impacts on agriculture and society, of regulations, and of public attitudes and public communication. Biologists on the Commission staff argued in the mid-70s for a Community research programme in biotechnology. At the time, most of our research effort was on nuclear power, under the Euratom treaty. It took 6 long years of argument to persuade our Member States to agree on what became the "Biomolecular Engineering Programme", (BEP) which ran from 1982-86, with a total budget of 15 MECUs.[1] That research and training programme, though small and limited to agriculture and agro-food topics, was highly effective in stimulating collaboration between the best laboratories in Europe. Between them, those labs have identified and characterised 20 distinct plant genes - and in total, there are less than a hundred known in the world today. They pioneered the use of the Ti plasmid in plant genetic engineering, including its extension to monocots. A new research programme, the Biotechnology Action Programme (BAP) was launched last year, at roughly treble the size of the first (55 MECUs). It will run from 1988-1991.

[1] The ECU is approximately comparable with the US dollar, subject to fluctuations of a few per cent either way.

But research by itself is not enough. Recognising the strategic need for a technology assessment capability, the Community in 1978 set up its own futures group, known as "FAST" (Forecasting and Assessment in Science and Technology). The first programme, from 1979 to 1983, produced as one of its outputs a *Community Strategy for Biotechnology in Europe*. This six-point plan was agreed by Commission and Council in 1983. It is described in the Commission publication COM (83) 672, *Biotechnology in the Community*).

## THE SIX-POINT STRATEGY

The key feature of this strategy was not just to back the research and training effort, but to get the message to all Commission services that biotechnology was going to be a major issue, and one that would require changes in many areas of policy. The six points are as follows: research and training, concertation of actions and policies, feedstock prices, regulatory regimes, intellectual property rights and demonstration projects.

Research I have already mentioned. Biotechnology is a knowledge-based activity, and Europe's efforts are often weakened by fragmentation and duplication of effort. Our research action programme in biotechnology is helping to overcome these problems, and building stronger links with industry, for as we are reminded by the first article of the new Title VI (Research and Technological Development) of the Single European Act.

> The Community's aim shall be to strengthen the scientific and technological basis of European industry and to encourage it to become more competitive at international level.

The research programme is also reinforcing the infrastructure for biotechnology, in terms of data banks and cell and organism collections. This area is of great importance to the competitivness of bio-industry in Europe; and in the same vein I should mention the work of the Task Force for Biotechnology Information, which advised the Commission on the development of Europe's information services in biotechnology.

Our concertation unit (CUBE) works in two dimensions: one being coordination with Member States, the other coordination between services within the Commission. For these various policies have to support one another. It is no use having brilliant technology, if raw materials like sugar and starch are over-priced for reasons of agricultural policy. Ten years' argument preceded the new regimes begun in March 1986. These are already having a significant beneficial effect on new investments.

On regulatory issues and on intellectual property and plant variety rights, our services are currently busy drafting and debating the need for new Community directives.The scientists are educating the lawyers and the lawyers the scientists, about their respective views of reality.

But rather than simply listing the points of our "strategy", which you can read in the various Commission documents, I should like to expand here on certain aspects relevant to the broader social dimensions of biotechnology.

## AGRO-INDUSTRIAL DEVELOPMENT, INFORMATIZATION AND DEMATERIALIZATION

The long battle over new price regimes for sugar and starch was just a small skirmish within the inevitable and continuing friction that is accompanying the centuries-long change in agricultural structures. It is clear that, barring catastrophes, the new technologies will enable agricultural productivity - and potential production - to continue growing faster than demand, for many years to come. Land prices are falling, the decline in the number of farms and farmers is continuing, even accelerating ; these are trends common to all reasonably organised economies. Competition for shares of world markets will bring supply into balance with demand.

So the productivity gains of biotechnology should raise quality and lower costs, to the general benefit of consumers and society. Jonathan Swift surely had biotechnology in mind when he wrote, in 1726:

> "... whoever could
> make two ears of corn or two blades of grass to grow,
> upon a spot of ground where only one grew before,
> would deserve better of mankind, and do more
> essential service to his country than the whole race of politicians put together".

However, now that we have got ten ears of corn where one grew before, the whole race of politicians is trying to work out where to store them, how to sell them, or what to do with them. This leads me to the subject of demonstration projects.

If we can solve the problems of adjustment, we must surely see the impact of biotechnology on agriculture as a major strategic opportunity. Science and technology have increased our range of options, and over the last couple of centuries have made possible both a huge expansion of "potential", because we do not always use wisely our freedom of choice; but the freedom, from disease and want, from ignorance and constraint, has certainly been expanded.

Biotechnology is removing the "Limits to Growth" (Meadows et al., 1972). But as the Club of Rome has also acknowledged, there are "No limits to learning" (Botkin, et al., 1979) The major achievement of biotechnology over the past 3 decades has been to demonstrate that mankind's key resource is organized information; DNA sequences or the design of a microchip, the essence is information. In the futures studies of the FAST group, we summarized a lot of our conclusions in two ugly words, which sound much more persuasive in French:

"INFORMATIZATION" and "DEMATERIALIZATION".

A nation's economic might is no longer to be measured in tonnes of steel or sulphuric acid produced - nor, perhaps, in millions of cultivable acres. Any analysis of sectoral statistics has to look at value, not tonnage; and look at the trend in value per tonne, or in value added per unit of input. Information workers already constitute 50 to 70 per cent of the workforce in a developed economy.

This is why at GATT, to-day's arguments are about free trade in services, about respect for intellectual property. Patent law, plant variety rights, the limitations which Defense and Commerce may try to impose on technology transfer - these are today's agenda, along with legal issues of copyright, privacy, and the ownership of one's own cells and DNA sequence. This is what the "information society" means.

Knowledge is a substitute for materials, a substitute for energy. We now spray grams and milligrams of pesticides, where yesterday we sprayed kilograms and tonnes; for today's pesticide can be more precise in the time when it is sprayed, in the size of droplet which carries it, in the design of its molecules, tailor-made to fit the receptors of the target pest. The same computer software is designing drugs to target cancer cells, hormones to modify metabolism. In brief, biotechnology is bringing the possibility of biorational control, of plants and animals, of their transformation into food and useful products, and of our own health.

Where does that leave the farmer? He has to add enough value to earn his living. That is increasingly difficult, for he is squeezed between rising input costs and falling prices. Every modern agricultural system is facing the social problems of structural adjustment, and there must be some support for the problems of these painful changes. If the farmer is to survive, he has either to accept the logic of scale and rationalization, or diversify and expand his range of activities; be it into environmental conservation, organic farming, specialist horticulture, or leisure activities, such as golf courses. We are looking for new ways of adding value to produce higher value molecules, and to produce outputs better adapted to the needs of the market.

The new opportunities demand collaboration between science, industry and agriculture. The Commission has stated its intention to launch a demonstration programme, using biotechnology to stimulate developments at the interface between industry and agriculture. Last year a formal invitation was issued, soliciting expressions of interest. Over 800 were received. From every Member State, from companies large and small, from agricultural research stations, from universities and individuals, the ideas have come pouring in; and some of them will be winners. (The Commission paper COM (86) 221, *Biotechnology in the Community: Stimulating Agro-Industrial Developmen*t gives an overview of the issues and possibilities).

## THE CONTEXT: REGULATORY ASPECTS, PUBLIC OPINION AND NATIONAL DIVERSITY.

Let's not get carried away. What I am foreseeing will be controversial when it is translated into specifics, and some of it will take a long time. What the public agencies have to provide are appropriate groundrules for enlargement and exercise of freedom, bearing in mind Rousseau's remark that, "my liberty ends where the liberty of my neighbour begins". That means regulation.

The Commission has stated its intentions (in the document COM (86) 573, "*A Community Framework for the Regulation of Biotechnology*"), to produce proposals on the regulation of biotechnology, in contained or non-enclosed use. We have, of course, many existing regulations for products which may be made by biotechnology: animal feeds, new plants, pharmaceuticals and veterinary medicinals, etc., and these regulations will partly or totally meet the need; the question is what needs to be added. We have participated in the OECD expert group on recombinant DNA safety, and we have consulted with our Member States. Our aim is naturally to achieve harmonised regulations for the whole European Community; but our basic treaties respect the rights of Member States to legislate more strictly where they feel public health or morals are concerned.

At present, there are significant differences of viewpoint, which reflect the diverse history, culture and experience of the different Member States. Consider the results of a public opinion survey carried out a few years ago, on attitudes to science and technology. It is not surprising, if these figures mean anything, that Denmark has been the first country to pass specific and apparently strict legislation on the use of genetically engineered ogranisms. The Netherlands also adopted a fairly strict view, in the debate which took place in the late 70s on laboratory work in rDNA; but their attitude today has evolved, as a result of Parliamentary enquiry:

> As for recombinany DNA activities, a DNA commission reported in August 1983 to the Dutch parliament that genetic engineering is controllable and acceptable. The parliament has used this advice to decide that there will be no special law or regulation promulgated for genetic engineering. The existing regulation will be maintained, but centralized "advice" pertaining to the risks of genetic engineering will take place.
>
> The Netherlands adheres to all safety regulations. The Dutch biotechnology industry considers Dutch regulations rather strict, but acceptable. Internationally Holland plays a leading role in formulating proposals to regulate the safest application of genetically application of genetically manipulated organisms. (Ministry of Economic Affairs, 1984)

It is not irrelevant to note that the quality of public information in the Netherlands on science matters of general concern, is outstanding; and on biotechnology there have been excellent posters for schools, readable non-specialist books, and a television series.

The Italians are favourable towards genetic research, for they know too well the consequences of genetic research such as thalassaemia, prevalent in the Mediterranean lands. But we should also bear in mind the controversy over single-cell protein, reflected in the other question on synthetic foods.

In Germany, the strong environmental movement finds expression not only in the Green party, but across the political spectrum, and this imposes caution on innovation; but many there recognize the potential of biotechnology for environmental improvement. The work of the Washington based Worldwatch Institute is admired, and Edward Wolf's paper on biological diversity is quoted;

> As these connections become better understood, a biotechnology attentive to natural history may provide some of the most powerful tools to reduce the pressures on genetic resources and enhance the value and conservation of wild species. (Wolf, 1985)

The potential of biotechnology for environmental improvement is recognised by the report published in January 1987 by the German Parliament's Committee of Enquiry into the Opportunities and Risks of Biotechnology (Bundestag, 1987). But it is a cautious document, and recommends a five year moratorium on the field release of genetically manipulated microorganisms.

The British have been active in biotechnology and rDNA regulation, as befits their leading historic role in the life sciences, and are confident that their understanding of the safety aspects enables them to be bold in innovation. The field release experiments in summer 1986 by the Institute of Virology in Oxford have been widely reported. They provide an admirable example of cautious and responsible experiment, of the type which is needed to provide the scientific base for regulation. We are glad to number them amongst the contractors in the risk assessment area of our biotechnology research programme.

## PUBLIC INFORMATION

Public confidence is a critical factor for biotechnology, and has to be based on trustworthy behaviour by the innovating industrialist or farmer, and by the regulatory authorities. This close relationship between regulatory policies and public information was underlined at the workshop, "Regulating industrial Risks", held at the Commission's Joint Research Centre, Ispra, in October 1984. The basic questions addressed were: "How can an appropriate balance be maintained between industrial progress based on technological innovation and the potential risk from these new developments to health and environmental well being? Can risk regulation processes be sufficiently effective and responsive to reconcile accelerating technological advances with growing pressures to avoid ill-effects form that progress?".

Amongst the key observations recommended by this workshop for consideration when developing specific policies, the following are particularly relevant to the present situation for initiatives in biotechnology regulation and public information:

3) Successful regulations cannot be based solely on scientific information, because scientific consensus appears not to be achievable and the regulatory process has to resolve social and polical conflicts that extend beyond scientific considerations. Furthermore the scientific community may contain divergent viewpoints and sometimes experts appear as advocates of a specific viewpoint.

4) Effective regulations must not only be scientifically sound, but must be practically implementable and command the respect of the organizations, groups and individuals affected.

5) The most effective "style" of regulation, in a particular context, takes account of regulatory experiences elsewhere and is adapted to the deep-rooted political and cultural conventions, as well as to the general administrative procedures in a particular country or region.

6) Communications media (TV, radio, press) are integral elements in political processes and, therefore, inevitably play a significant role in shaping regulations, in the allocation of resources to regulatory institutions and risk research, and in

influencing public support for or against regulations. Policymakers must, therefore, give adequate attention to the media. (Otway and Peltu, 1986)

On some issues where scientists and industrialists have been annoyed by regulatory stances they reckoned too strict, but one must remember that the members of the public and Parliamentarians are concerned less with theoretical risk analyses than with empirical fact; and in particular with enforceability. Scientific advice is important, but it is not decisive; what is decisive is public trust in the regulatory authority. Amongst other things that implies a certain openness about how standards are set and a willingness to be seen to act.

On the other hand, if regulation is introduced which is unrealistically strict, unenforceable or irrational, then the loss of confidence by the professional circles concerned risks the development of illegal activity and the regulatory authority losing respect.

The Commission has launched some initiatives in the area of public information, for, as Bernard Dixon has well expressed it; "public clamour over a vast range of topics, from nuclear power to recombinant DNA manipulation, has moved far ahead of scientific literacy in the population at large." (Dixon, 1986)

The difficult thing to communicate in biotechnology is the "package deal" which it offers. There are four key characteristics of biotechnology; and they are closely related to one another:

i) the inherent scope for futher scientific discovery, and hence vast but unpredictable technological potential, which has been opened up by the cumulative (and recently accelerating) progress in the life sciences;

ii) the unremitting competitive economic challenge, agricultural and industrial, between the major blocs and between the major companies, in which the use of advanced technology (including bio-) is of growing importance;

iii) the (justified) hopes of reducing or even eliminating hunger and disease;

iv) widespread popular apprehension about non-understood science and socially or culturally unacceptable innovation.

## THE RISKS OF IGNORANCE AND THE COSTS OF DELAY

Implicit in the above four key characteristics are associated risks and costs. Ignorance and misunderstanding leads to inappropriate responses, by individuals and societies, to the challenges and opportunities of biotechnology. If we do not regulate, we may expose ourselves to dangers which become apparent only after significant harmful effects have become visible; if we over-regulate, ban or delay, we lose the beneficial effects of potential innovations. How does a society strike a balance between these costs, under conditions which always include partial ignorance ?

We have to recognise that we are always in a learning situation, and regulations need to be adaptable enough to evolve with advances in understanding; indeed from the start, they need to permit enough experimentation to advance understanding. The balance of judgement vis-à-vis the new technologies of genetic engineering and other aspects of biotechnology has to be overwhelmingly in favour of rapid innovation. Consider the two major application areas: health care and agriculture and the environment. The costs of delay are in both cases enormous.

In health care, a hundred thousand people are dying every day, from starvation and preventable or treatable diseases. Many more are suffering from chronic ill-health and malnutrition. We can through biotechnology reduce the costs of food production and health care. For example, through these new techniques we can study the constituent parts of viruses, and manufacture in large quantities and high purity their antigenic proteins. The development of monoclonal antibodies is providing precise and effective diagnostic tools, and the basis for delivering drugs to precisely targetted cells in the body. Such technologies are essential in the continuing battle against parasitic diseases such as Chagas', or to combat the continuing and ever-changing challenges of malaria. They are equally needed to confront novel challenges such as the AIDS virus.

There is much talk about the hypothetical risks of intervention in the human genome. Nobody who has seen a child dying over many years from a genetic disease such as muscular dystrophy would hesitate to support continued and bold research, to develop understanding and the means of intervention in such cases. Many societies carry heavy burdens of genetic disease, which blight individual lives. Only in exceptional cases can it honestly be claimed that the suffering is sublimated into some higher spiritual or philosophical achievements.

Concern is sometimes expressed about the impact of biotechnology on agriculture and the environment. Around the world, some two hundred million people are engaged in "slash and burn" agriculture. Of course, they are driven by their local necessities; but by their actions, they are destroying the remaining areas of tropical forest, with a loss of species estimated at several hundred per day. We are likely to lose half of all current species within twenty years; it is an environmental catastrophe, a species extinction of a magnitude unparalleled since the death of the dinosaurs. It is not hypothetical; it is happening now. Yet biotechnology can offer the means to feed the world's whole population more than adequately, using far less land than we cultivate today. Further major gains in agricultural productivity are now seen to be possible. By pursuing these, we should be able to take the pressure off the environmentally sensitive areas, such as the uplands and the wetlands, the unique ecological habitats. We should be able to defend our forest, be it in temperate or in tropical zones, and restore our environment, replanting appropriate species in degraded areas, such as parts of our Mediterranean littoral here in Europe.

The political and economic difficulties should not be understated. But it is essential that interest groups seriously concerned with the maintenance and restoration or enhancement of the environment, of ecosystems, should recognise the potential of the recent advances in biotechnology for enabling their objectives to be achieved.

## PUBLIC CONCERN AND SCIENTIFIC METHOD

There is in some countries and in some interest groups an upsurge of public disquiet about the implications of progress in the life sciences, and their applications. The reasons are less a fear of specific consequences, than a fear of the unknown, and a facile transfer of concerns appropriate to one specific danger to other areas in which they may be irrelevant.

Such sloppy thinking is dangerous; it costs lives. As the Botkin report puts it, in the absence of widespread understanding, we "shy at kittens, and cuddle tigers". (Botkin et al., 1979). There is more damage to human DNA from tobacco smoke than is ever likely to result from rDNA laboratory work.

> Tobacco causes more death and suffering among adults than any other toxic materials in the environment ... involuntary exposure to cigarette smoke causes more cancer deaths than any other pollutant. (Chandler, 1986)

Issues are not resolved by statistical analysis, however, but by public confidence; as Harry Otway has expressed it: "if the public cannot evaluate the risk, they will evaluate the regulator". Confidence has to be built upon trust, and whatever the short-term aberrations, the trust has ultimately to be built upon regulations and practical experience which are scientifically based and objectively evaluated.

The scientific method is one of the supreme achievements of human culture over the past five centuries. It is the direct lineal descendant of Renaissance humanism, in its self-reliant dependence on human observation and deduction. It is the most effective societal learning instrument ever developed. It is naturally international in character.

Key features of science are its cumulative character - we build on the achievements of our predecessors - and its openness to correction. In spite of its achievements, this openness to correction implies an essential humility, which Karl Popper expressed in his phrase, "the conjectural status of knowledge". All scientific knowledge is hypothesis, subject to refutation or refinement by new evidence or experiment. Thus to refuse the further expansion of science, or to be nervous of its application to useful purposes, implies a failure of self-confidence in human capacities, and of confidence in our societies' abilities to learn, to correct and to improve.

Biotechnology is important for human progress, and for the management of the global ecosystem, a responsibility which we cannot escape or "leave to nature". Public understanding, attitudes and acceptance will increasingly be of strategic significance for the progress of biotechnology. Public education/information is therefore of strategic importance. It is arguably, the single most important strategic instrument in a society's self-management, learning and survival, whether we define "society" in global, regional or national terms. This is particularly true of an area like biotechnology, which touches so many areas of social life.

In conclusion, I would like to congratulate the Dublin Foundation for organizing and preparing this forum, and for the studies they have been supporting over the previous months. These have a very practical value; for we shall over the next twelve months be reviewing and extending the strategy developed five years ago, and trying to foresee and prepare for the Community role in biotechnology in the mid 1990s.

In this process of reviewing our strategy for biotechnology, we shall undoubtedly place greater emphasis on the social dimensions, and on information of every sort -

ranging from the global infrastructure for scientific information, to information designed to raise the quality of political debate.

I have no doubt that occasions such as this are of high value in drawing attention both to the marvels of biotechnology, and to the challenges of its political management. I think that the open societies of the Western democracies have two strategic advantages in biotechnology. Firstly, our open flow of information stimulates our science, in both cooperative and competitive activities. Secondly, our open processes of challenge and enquiry enable us to control the new technologies in the service of human needs.

# 3 Historical Perspectives on Biotechnology

Edward Yoxen

## DOMESTIC ORIGINS

The special characteristics of biotechnology derive from the interaction of the old and the new. In particular it is the intensification of interest in research and development, which Jacobsson and colleagues call the scientification of production, that merits our attention (Jacobsson et al., 1986, p.2). We cannot understand the dynamics of events in the present unless we also have a picture of past trends, interactions, and conflicts. That offered here will be far from complete and heavily skewed towards events in the UK and the US, but nonetheless it illustrates the important stages of development and the questions they pose.

The oldest forms of biotechnological transformation were practised in the household. These included the production of alcoholic drinks from fruits, vegetables and grains, the fermentation of milk and its derivatives to make cheese, yoghurt, buttermilk, and even alcohol, the fermentation of bean curd and similar substances to make tasty, protein-rich foods and sauces, the tenderising and preservation of foods by wrapping them in leaves, the use of yeast to make bread dough rise and the composting of household wastes. In many parts of the world such procedures still form part of the daily household labour of millions of people, most of them women. Recipe-books, herbals, handbooks of domestic practice are thus an archive of early biotechnology. Drawing on such sources Barker has recently described household methods of preservation, brewing and cosmetic production from the early 19th century. (Barker, 1986)

Some of these practices, along with the use of herbs in traditional remedies and simple procedures for crop protection or fertilisation in peasant agriculture, are now the object of intense scientific scrutiny. Ethnology has become another mode of invention, capitalising accumulated local knowledge by testing foods and remedies, isolating active ingredients, optimising particular processes and, in some cases, transferring them to a completely different technical setting. The biotechnological expertise of less developed countries represents a reservoir of invention, long since destroyed in the industrialised world. How it should be valued economically is a moot question.

One of the myriad effects of industrialisation, with major changes in patterns of consumption, has been the transfer of such activities from the household, via local artisanal enterprises, into the modern factory. The processing of biological materials has changed from being the exercise of domestic skill using rudimentary technologies, like a still or a churn, to an industrial transformation, carried out in huge volumes, and requiring the most advanced scientific input for its controlled operation. The precursors of such science-based processes survive nonetheless in the industrialised nations, in the luxury 'homemade' substitutions for standardised products, which sustain a small market for ingredients produced by biotechnology: dried yeast culture for bread-making, home brewing kits for beer and wine production, and pectinases for jam-making. These products are of course scientifically formulated to last long enough in the modern retail system to reach the consumer. Biotechnology gives us back a version of the domestic past.

## SCIENTIFIC CONSULTANCY FOR INDUSTRIAL MODERNISATION

By the nineteenth century the industrialisation of biotechnological production had gone some way in brewing, as local firms gradually began to increase their sales in major European cities and to extend their commercial dominance over the surrounding regions. But problems of process reliability and the restricted length of time for which beer could be kept represented real constraints on profits and business expansion. In Britain brewing was limited to the cooler months of the year. In other European countries major investments were made in cellars and ice-stores to try to keep a less rapidly fermented and generally lighter beer in good condition. Some innovations, such as the use of a square stone fermentation vessel which could be scrubbed clean and from which air could be more easily excluded, show that brewers were prepared to experiment with new ideas, despite the strength of craft traditions, but without the benefit of scientific advice. (Sigsworth, 1965)

By mid-century a few of the larger and more progressive companies were buying the services of scientific consultants, the most famous of which was Louis Pasteur. His crystallographic interest in the forms of tartaric acid led him to investigate its production in various industrial processes, which in turn led to an interest in fermentation generally. One of Pasteur's contributions was to promote the view that fermentation be thought of as a biological process, effected by microscopic living organisms that transformed a substrate chemically as they grew. He also believed that this occurred most efficiently when oxygen was excluded. For him fermentation was 'life without air'. These evolving, and to some extent erroneous, views he put to work in a series of classic reviews of industrial processes, in the vinegar industry, in the brewing of beer, in wine production, and, less successfully, in the silk industry.

These works appeared in the 1860s and 1870s, before he began his medical investigations, and were welcomed by the more research-oriented brewery owners throughout Europe. They offered new ways of thinking about the contamination and spoilage of beer and wine, as the result of microbial infection, and a new method of preserving milk, wine and beer by heat treatment. They demonstrated what chemical and microbiological studies could offer the fermentation industries, and several companies set up laboratories that have since become major research centres, such as the Carlsberg Laboratory in Copenhagen, which later developed expertise in plant breeding. (Holler and Moller, 1976) Biotechnologists today often regard Pasteur's ideas about fermentation as marking a historic transition, even though in some respects they were quickly modified by other applied scientists. Hansen, for example, in Copenhagen showed that contamination could be reduced by using pure yeast strains and excluding wild yeasts from the fermentation. (Teich, 1983; Peckham, 1986). Interest in research in the brewing sector stemmed from and reinforced the trend to economic concentration, which had important effects in the reorganisation of primary production in agriculture. (Weir, 1984)

Studies of fermentation also led to the idea that reactions in living cells were catalysed by substances of great specificity, that came to be called 'enzymes'. This theory was very important for the nascent discipline of biochemistry, as it became differentiated from physiology. (Kohler, 1982) By the early 20th century fermentation processes were of some significance to the expanding chemical industry. A famous example is Chaim Weizmann's research, begun in Manchester, on the production of acetone and butyl alcohol from starch, using a bacterial culture. This was very important in the First World War, as British explosives manufacturers had depended on supplies of acetone from Germany and Weizmann was funded by the government to develop his process.

(Rose, 1987)

Also processes were developed to produce industrial alcohol using surplus molasses from cane sugar refining. (Hastings, 1971) One use for it was as a fuel additive. But many of these processes became uncompetitive with oil-based syntheses, and to a large extent that remains the case today. There is some disagreement amongst economic historians as to the rate at which this occurred. Haber's account of events in the United Kingdom is given in the next section. In the brewing industry in the 1930s the first attempts were made to switch from batch to continuous production. Even though the technical problems were eventually overcome by the 1950s, the seasonality of beer sales made it uneconomic. (Bud, 1984)

As plant size increased to reap economies of scale, and new processes with greater efficiency, changes followed in working practices and in labour productivity. A fascinating example of this is to be found, in a very different geographical and cultural setting, in the history of the Kikkoman Corporation in Japan, the oldest continuously operating business enterprise in that country, and the largest producer of soy sauce in the world.

Several points can be abstracted from Fruin's remarkable history of the company. (Fruin, 1983) Firstly in the 19th century and earlier the fermentation of soy sauce was a highly traditional craft activity, carried out in small, family-owned enterprises. The city of Noda, near Tokyo, was an important centre of production. But those who owned the business often played a minimal role in its operation at any level, and certainly not in production. Secondly towards the end of the 19th century, as the country began to industrialise, the structure of the soy industry began to change, with the formation of cartels of producers, often linked by ties of kinship. This led to the appearance of new kinds of business organisation, to economic concentration and to incentives both to develop new process technology and to restructure the work force. Thirdly the attempts in the early 1920s to change from the traditional system of recruitment, payment and supervision based on contracted work-gangs to a more modern employment system, which required more intensive work at a regular pace from labourers and greater technical supervision by the labour contractors, led to increasing unrest and a historic series of strikes in 1923 and 1927-8.

The issue was never one of technological displacement of workers in the sense of layoffs and dismissals. It meant discomfort, possible job dislocation, and probable disorientation. The entire structure and rhythm of work as well as the work-place were being altered. Although the introduction of boilers, hydraulic presses and mechanical mills did not yet change the sequence and number of stages in the production process, the volume and rate of throughput was greatly accelerated. The potential for greater production was introduced; whether or not it would be realized depended on the cooperation and motivation of the workforce.

> Workers were troubled by the prospect. Toji, the traditional sources of legitimacy and authority in the factories, were either elevated and made into factory managers, or else they were supplanted by managers who were sent into the plants as the lowest rung of a centralized managerial hierarchy. The latter represented authority, but they knew little about soy sauce brewing technology and even less about industrial relations on the shop floor. (Fruin, 1983, p. 170)

Fruin indicates that one important result of the 6 years of conflict was the formation of an enterprise union, of the kind that is now regarded as typical for Japan, and a prolonged attempt by management to promote an ideology of 'the firm as family' to promote acceptance of new working conditions and a new management structure.

## STRATEGIC DEVELOPMENT UNDER THE STIMULUS OF WORLD WAR

In the late 1930s Fleming's work on the antibiotic action of penicillin moulds was re-examined by a group of biochemists in Oxford and procedures worked out for identifying and extracting the substance involved. It soon became clear that scaling-up these promising laboratory results would require an enormous effort. This was undertaken first by pharmaceutical companies in Britain, in ways that required a prodigious amount of labour, to harvest the hundreds of thousands of cultures growing in milk bottles. This effort was then passed on terms that remain controversial today to firms in the US, who were able to draw on the experience of the US Department of Agriculture in growing bacteria on dairy wastes. Liebenau's comparison of moves made by the companies involved in Britain and the US highlights the fact that the Americans tended to take a longer term view, and to plan for the development and supply of a major post-war market. (Liebenau, 1984)

The American effort led to radical process innovation, with so-called deep fermentation processes, which successfully produced penicillins in commercial quantities, and at prices that fell rapidly as the process was refined and more productive organisms discovered. This work is often seen today as a second major milestone in biotechnology, after Pasteur's studies almost a century earlier. (Houwink, 1984) It led to a new phase of research-driven innovation for pharmaceutical companies around the world. This has produced a tremendous increase in their sales, their size as companies, their expenditure on research and on promotion, and in the pressure to recoup very high development and market-building costs before the expiry of patent cover on products causes profits to be rapidly competed away. This situation has been exacerbated since the 1960s with increasingly stringent regulation and more effective product imitation.

The growth in pharmaceutical sales and increased expenditure on biomedical research since 1945 has also stimulated much research in bacterial genetics and physiology, to discover how such organisms grew, how high-yielding strains could be selected and how micro-organisms can become resistant to antibiotics and can transfer this trait between different strains. This led for example to the discovery of bacterial plasmids, which have assumed an enormous practical importance in the genetic engineering of the late 1970s and 1980s.

In another area too it has been claimed that the 1939-45 war brought about developments of major importance for biotechnology, but this time in the opposite direction. Haber has suggested that the UK chemical industry was relatively slow in becoming a petrochemical industry. (Haber, 1980) Thus in the 1930s substances like benzene and other aromatics were derived from coal-tar, and little interest was shown in obtaining them from petroleum fractions. Similarly the alcohols, and their primary derivatives, such as acetone and acetic acid, were obtained by fermentation. Indeed ethanol production was protected by tariffs, much to the benefit of major producers, such as Distillers. With the advent of war, it was realised that using molasses from overseas as a fermentation substrate had very high opportunity costs as space on shipping across the Atlantic was so scarce. Two UK companies began experimenting with petrochemical feedstocks, but only as an expedient in the exigencies of war. Whereas in the United States and elsewhere the use of oil fractions instead of coal, or calcium carbide or fermentable substrates as feedstocks declined rapidly in the 1940s, use of these materials persisted in Britain into the 1960s.

## POST-WAR EXPANSION

One of the characteristics of the 20 year period of economic growth since 1945 in the industrialised nations has been the successful realisation of economies of scale, particularly in industries based on flow processes, such as chemicals, oil refining, pharmaceuticals and food processing. With the switch to oil products as a feedstock in the chemical industry came the construction of basic plant on an ever increasing scale. This had many effects. One was to diminish the flexibility of production. To recoup the cost of building the plant, it had to be run at peak efficiency for as long as possible and its output sold. Another was to consolidate a managerial ideology of bulk production from highly engineered process plant. In the 1960s this influenced the plans for huge single cell protein plants, like the one eventually operated by ICI in the early 1980s. By then the commercial and technological environment had changed, but within companies like ICI some people found it hard to see that biotechnology had any immediate potential, given that its markets were likely to be relatively small.

Another effect was to expand the market for the services of process plant contractors.

This industrial sector has been examined by Freeman and his colleagues in the 1960s, and more recently by Barna. (Freeman et al., 1968; Barna, 1983) Their common concern has been to discover to what extent such contractors contribute to innovation, either in the costly business of originating new processes or in assisting the diffusion of others' work by building plant using licensed technology.

Barna writes:

> In the UK contractors have drawn heavily on research done by firms such as ICI, British Gas and the British Steel Corporation. ICI has followed a fairly liberal policy, granting licences to a select group of contractors rather than a single one. ... West German contractors tend to carry out more research and development themselves and this is explained, at least in part, by the restrictive licensing policy of the German chemical industry. (Barna, 1983, pp. 174-5)

The implication is that contractors in West Germany have to be more active in originating new designs and new processes to stay in business. This may well explain the important role played by the German process plant manufacturers' association DECHEMA in getting their government to formulate a plan for biotechnology, prior to the appearance of recombinant genetics, in the early 1970s. (Jasanoff, 1985)

Finally it is obvious that an industry so dependent on oil as an energy source and as a feedstock and organised around the manufacture and sale of bulk chemicals would be particularly vulnerable to increases in the price of oil, of the kind that occurred in the mid and late 1970s, and would be faced with major rationalisation when the output from its huge plants could not be sold at a profit in a period of recession. One part of that rationalisation has been a change in the organisation of R and D, in some cases at least away from the very large university-like research laboratories that companies like ICI funded in the 1960s. This has been one factor that assisted the growth of research-intensive new biotechnology firms as specialised research contractors in the 1980s. (Kenney, 1986)

Two other features of this period of economic expansion require very brief mention. The first is the intensification of agriculture, and the development of its capital- and energy-intensity, assisted by a significant degree of government subsidy, intended to stabilise farm incomes, to create a secure food supply and to encourage innovation. For the chemical industry the agricultural sector has a provided an important market for fertiliser, pesticides and herbicides, all of which require considerable energy to produce and to spread on the farm. Whilst farmers have in general been able to prosper under such conditions, a decreasing proportion of the value added to agricultural products has been generated on the farm. Although the price of agricultural commodities has been often been maintained at artificially high levels, both the food processing and food retailing industries have also profited from continuing agricultural modernisation.

Moreover the output of the food industry is enormous. Volumes of production often exceed those in the chemical industry. For example Dunnill demonstrates that world production of milk and of sugar exceed that of naptha (an important chemical feedstock) and of aluminium. (Dunnill, 1981, p.205) Even so in recent years governments have been urged by the food industry to couple agricultural research more directly to the needs of the processors and retailers and less to those of the primary producers.

The second area that must be mentioned is health care and the expenditure on the products of the pharmaceutical industry. The industrialised nations now spend around 8-12 percent of GNP on health care. Of this the largest proportion goes on labour costs. In some countries the health care system is the largest employer. Nonetheless around 15-20 per cent is spent on drugs and medical supplies, which represents a very sizable market for the major firms that dominate the industry. Despite the costs of developing and marketing new products the return on capital employed in this sector has been significantly above the average for industry as a whole. Nonetheless governments have sought continually, but with varying degrees of commitment and success, to control drug pricing. The pressure to do this has increased in the last decade as major attempts have been made to reduce public expenditure.

On the other hand the post-war period has also seen a marked increase in expenditure on fundamental biological research, most strikingly in the United States, with the growth of the National Institutes of Health. Historians such as Stephen Strickland and others have argued that this expenditure on research was a covert way of supporting medical education and modernising medical practice in the face of lobbying from the American Medical Association to maintain the conditions under which doctors could charge high prices for their medical services. (Strickland, 1972) In effect the State created more medical researchers rather than more doctors. Those students who went on to become physicians were trained in research-oriented medical schools. Clearly such an explanation could not be applied without significant modification to European countries. However it is clear that here too doctors have been able to retain their authority over an increasing number of supporting professionals. Similarly many more people have found the idea of a research career in biology more appealing than was the case before 1939. One result has been the creation of an enormous base of knowledge and technical skills, on which industry can now draw, and the formation of a highly competitive research community, whose members must now turn increasingly to industry for research funds. It is only in very recent years that either party - industry or research biologists as a group - has taken that prospect seriously.

## STRUCTURAL TRANSFORMATION DURING RECESSION

Manifestly in the 1970s many of the operating principles of post-war industrial production began to lose their effectiveness, as energy costs rose, markets saturated or contracted, and new competitors appeared in areas like textiles, shipbuilding, electronics, steel and car manufacture. The most dramatic result was a steady growth in the number of people employed. Throughout the industrialised world governments began to practise stricter financial control, cutting public expenditure whilst reinforcing fiscal incentives to innovate. In the US changes in tax laws encouraged the growth of venture capital funds for the support of new technology-based concerns. As the recession deepened in the 1980s many companies were forced into major changes of business and management strategy, getting out of relatively unprofitable areas of production altogether with wholesale plant closures, trading factories and facilities with competitors, making major acquisitions to take them into new areas, restructuring internally, to achieve greater flexibility in the use of the factors of production, and examining the effectiveness of research and development.

The evolution of biotechnology is connected with this complex process of industrial transformation in many different ways. The principal purpose in reviewing the history of early forms of biotechnology is to remind ourselves of the industrial context within which important innovations of the mid 1970s have had an appeal. It is the way in which such techniques have been taken up by companies within several sectors of production that is important. To put that another way, biotechnology only exists as research, process plant, products and services shaped for particular industries and their characteristic markets.

Thus many of the important scientific discoveries and inventions, in particular the development of immobilised enzymes, the culture of plant tissue *in vitro* and the regeneration of plantlets, the construction of electrodes that measure important biochemical parameters in real-time, the use of computers to monitor fermentation processes, the discovery of the techniques for the construction, movement and replication recombinant DNA molecules, and the fusion and culture of antibody- producing cells, predate this period of radical structural change in industry. Nonetheless its occurrence had a profound effect on the rapidity with which and the means by which such ideas were commercialised.

It soon became apparent, for example, to scientists, to banks and finance houses, to corporations and to governments, that a global market for research and development services existed. New businesses created to commercialise research ideas based on recombinant genetics or monoclonal antibodies or novel bioprocess technology could expect a reasonable chance of survival, and might indeed be an extremely profitable investment. They could subsidise their own proprietary product development and the recruitment of specialist personnel and the construction of expensive facilities by contracting for others.

At the same time the 'new biotechnology firms' were not the only kinds of concerns taking an active interest in the field. Hacking mentions three other kinds, excluding the process plant consultancies and contractors and the laboratory equipment manufacturers. (Hacking, 1986) These are firstly Japanese fermented food manufacturers, notably Ajinomoto, who have moved into fermentation, first of amino acids and then of antibiotics, as extensions to their business. In general European and American businesses based on craft fermentation have not diversified in this way, the Dutch firm Gist Brocades being an important exception, as an enzyme supplier. Secondly pharmaceutical and chemical companies have become seriously involved with biotechnology, despite the fact that fine chemicals and antibiotics form only a small proportion of the bulk business of the chemical sector. Thirdly there are those companies who have traditionally been involved with the processing of a potential substrate. These include the American corn wet milling companies, who make corn oil, starch and animal feeds. This would also include companies involved in sugar refining, trying to find new uses for sugar.

Thus whilst venture capital-backed businesses may have had a striking rate of growth in the United States, particularly in the 1980s, it would be wrong imagine they alone constitute the industry. It makes more sense to see them as just one means of effecting the 'scientification' of production processes based on applied biology, in a period of change during which established firms may choose to contract out some of the risks of development. This applies not just to the innovation of new products, such as biologicals synthesised in recombinant micro- organisms, but also to fermentation scale-up and process design. (Van Brunt, 1986) Commentators like Kenney are obviously right to point to the importance of entrepreneurship in the diffusion of new forms of biotechnology, but of equal importance are the underlying structural conditions that make such activity likely to have an effect.(Kenney, 1986)

This suggests that it would be worthwhile to consider first some of the general economic and organisational objectives being pursued by companies at the present time and then to ask to what extent an involvement with biotechnology is seen as serving such goals. Thus for example chemical companies are seeking to minimise some of the competitive pressures upon them by finding niche markets for higher value, more specialised products. Not all of these can or will be made biotechnologically, but many of the novel products derived from recombinant genetics and those being refined by protein engineering fit this pattern. As Cantley reminds us in his contribution to this volume, Sargeant has described the historical shift to such commodities as 'dematerialisation' of production, as products such as pharmaceuticals, pesticides and industrial enzymes become more sophisticated, more fully the result of a design process, more powerful and specific in their action. The inevitable consequence is that they will be sold in smaller volumes. (Sargeant, 1984)

Another important general trend, much discussed in the economic literature, is a search for flexibility, although it is clear immediately that this may mean very different things in different industries. Some writers see it as enabled by new technologies, such as those associated with micro-electronics and as such fundamental to a new phase of economic expansion. Carlota Perez writes of the emergence and diffusion of 'the new productive common sense' as follows:

> ...[A] very salient characteristic of the new technological system is its capacity to cope with variety, diversity and dispersion at all levels, as opposed to the prevailing need for 'massification', homogenization and agglomeration typical of the paradigm about to be replaced. (Perez, 1983, p. 360)

Perez belongs to a group of economists committed to a long term cyclical view of the world economy, in which phases of expansion, stemming from the particular opportunities afforded by a key factor of production, come to an end as market conditions alter. Growth only resumes after major changes in the structure of the economy and within industrial and research institutions.

Other writers, Kaplinsky for example, have reacted against a kind of determinism in models of this kind, and have stressed the importance of the shift from flexible, small scale to mass production in the late 19th century and in the reverse direction in the 1970s and 1980s, in creating relations of production that are most easily managed, in the conditions of the time. (Kaplinksy, 1987) How this applies to flow-process industries is not entirely clear. The example quoted above of changing technology, shop-floor hierarchies and wage systems in the soy sauce industry seems to follow a trend towards larger-scale, less flexible production earlier this century. Much contemporary writing about biotechnology seems blithely to assume that the organisation of production is almost an irrelevance, except perhaps for problems of purification and workplace safety. But even the most capital-intensive plant has to be designed, built, maintained and its output distributed. This suggests that examining how the most recent forms of biotechnology are intended to modify, and actually have the effect of modifying, the organisation of production in various sectors would be a very useful exercise.

## AGENDAS

In this final section I want to sketch some possible research agendas covering various aspects of biotechnology. No attempt is made to be exhaustive, partly for reasons of space. I have tried to present the topics as connected parts of a whole, since it does seem essential, when given the pervasiveness and generality of biotechnology, to bear in mind that the likelihood and extent of change in one area of social life or sphere of production is often conditioned by developments elsewhere, also associated with biotechnological innovation. Thus for example the impact of biotechnology on chemical production depends partly on the future of agriculture. If agricultural materials were to be available much more cheaply, then chemical companies would be interested in changing those processes to use them to produce feedstocks. Similarly if there is a shift to less intensive farming in Europe, as one way of coping with the problem of agricultural over-supply of commodities like milk, grain and sugar, then the demand for many bulk chemical products will fall, and manufacturers will face even stronger pressures to diversify, for example to produce 'smarter' products for agriculture. Yet again another area for diversification is into health care products, which implies some impact on medicine.

### Changing conditions in agriculture

This area is one where it is essential to speak very cautiously of biotechnology as a cause of change. Farm production, the sale of agricultural commodities, and the processing and distribution of the resulting food products, form a very complex system, reinforced in Europe by the Common Agricultural Policy. Various forms of biotechnology, such bovine growth hormone, new veterinary products, pesticide-resistant crop varieties, new forms of slurry digester and so on, will soon allow marginal increases of efficiency and productivity, but their effects are conditioned by the operation of the system of production as a whole.

Two recent studies suggest how this area could be approached. Duchene, Szczepanik and Legg have produced a very detailed analysis of the problems in Europe created by the 'success' of the Common Agricultural Policy by the early 1980s. (Duchene et al., 1985) They describe the evolution of policies at the European level intended to secure farm incomes, particularly in those areas of the original Community of Six, where agricultural production was small-scale and backward. The system of guaranteed prices and subsidies for modernisation and diversification proved very effective, particularly when exploited by larger, more efficient producers, with the result that the enlarged Community, (10 when this study was written, now 12) is a significant exporter of food products on to a world market, where new economic and political rules apply. This is true even though in certain areas, particularly animal feedstuffs, forestry products and tropical food produce, the Community is also a major importer.

As is well known the need to reform the CAP is now very pressing. One of the problems is that producers often represent very powerful political lobbies, willing and able to put pressure on national governments when their economic interests are threatened. Another is that some reforms would have much more serious consequences for those countries with small-scale, less capital-intensive farms. One question then is how to stimulate productivity in areas for which there is a demand for the goods concerned and to encourage diversification amongst farmers. From within the Commission have come proposals to subsidise pilot projects in agricultural biotechnology across a very wide range, but which have as their goal a modernised, more highly market-driven agricultural system in Europe. (CUBE, 1986) It is acknowledged though that a shift to new methods of farming, new crops, new forms of land use, may bring environmental effects that could be problematic. Such effects are linked to and partially caused by economic transformations.

> It is already clear that biotechnology, applied in the field of agriculture, has the potential for both adverse and beneficial consequences for the environment.... The resulting opening up of new markets and enterprises through biotechnology should be such as to provide the farmer an economic incentive ... to maintain and enhance the rural fabric.
>
> On the other hand ... there could be;
>
> 1) changes in farm structure and employment;
>
> 2) increased pressure for industries to buy and operate farmlands for the production of specialized raw material for their industrial operations (with associated questions of soil quality and erosion);
>
> 3) a possible tendency in some regions toward monocultures centered on industrial needs, that is large areas of single-crop production, no fallow lands, and ensuing implications for the use of agricultural chemicals;
>
> 4) unforeseen consequences:
>
> - those linked to the deliberate release of modified organisms into the environment;
>
> - water and air pollution dangers and waste management requirments;
>
> - impacts on wildlife, habitats and ecological diversity. (CUBE, 1986, pp. 10-11)

The question of the impact of changes in technology on farm structure and employment has been the subject of inquiry in Europe and the United States. For example Bijman and colleagues, in their report for the European Foundation, estimate possible effects on farm employment in beet sugar production. (Bijman, 1986) Similarly in the United States the Office of Technology Assessment produced a major survey of trends in American farming, concluding that by the year 2,000 it was possible that 75 per cent of US production would come from very large farms, a long way from the traditional model of family businesses. (OTA, 1986) Medium scale farming will face constant pressure to survive over this period, whilst small scale farming will only exist at the extremes of rural poverty. There is obviously scope for a great deal more work here, taking into account the diversity of contexts within which various aspects of biotechnology are likely to play a role.

Reconfiguration in the process industries

In the late 1970s much of the political debate about biotechnology concerned the means by which laboratory workplaces and factories with scaled up production plant could be made safe, to the best practicable extent, given the potential risks to health that recombinant micro-organisms were thought to pose. (Bennett et al., 1986) Impact on working conditions in that sense remains an important issue, even if scientists' attitudes to this question have changed. It is important for example to ensure that safety regulations are applied in new biotechnology firms. It would be wrong to assume that just because enterprises are new and science-based that their workplaces are safe. There is scope for study here of how such new businesses develop internal systems of training, supervision and management review, given that they are usually built from scratch under considerable financial pressure with rapidly changing technology.

But there is another much more pervasive sense in which working conditions are changing, as industrial sectors are restructured. The basic purpose of the long historical section at the centre of this paper was to suggest that biotechnology be seen as part of this reconfiguration of industry, in which access to research expertise is vital and where new processes and modes of manufacture are being designed and implemented. As a consequence company structure and ways of working are changing.

As mentioned earlier there is a growing economic literature that claims to discern a trend in industry away from scale enterprises that form part of a network between the nodes of which production can be switched as economic circumstances alter. In some industries there is apparently a trend towards the creation of a set of dependent, external suppliers, with important changes in the size and structure of the work force of the dominant firms. (Murray, 1983) Linked to this a move towards internal company 'flexibility' in the use of labour, with marked differences in working conditions, job tenure, and payment systems between new categories of core and peripheral employees. Again as stated earlier there seem to be few studies that have examined structural transformations in the process industries, although as one technician in the major research complex of a food company put it to me, 'If your group is not into biotechnology, your job's on the line.'

Health care

This field covers a very wide range of economic, political, institutional and moral questions. Although we tend not to think initially of health care in these terms, health care goods and services are produced by labour. As noted already the health care system is a major employer, particularly of women workers. Historically as a service industry it has seen relatively slow increases in labour productivity, compared with manufacturing industry, and all European countries now face serious economic and political problems in attempting to control high hospital costs. In the United Kingdom attempts have been made introduce financial accounting and decision-making from industry for this reason. Also successive Conservative governments have sought to persuade more people to buy their own health care services from private suppliers outside the nationalised hospital system. Against this background there are two kinds of issues to consider: firstly whether new products and services will in fact be taken up, given the pressure on budgets and lack of time to test and assimilate new technical possibilities; and secondly whether particular techniques, such as the antenatal diagnosis of genetic disease or the early detection of some forms of cancer by new diagnostic kits, will have a major impact, given the moral problems that they raise.

Thus Farrands, for example, in a rapid but interesting survey of new technologies in medicine writes:

> The general practitioner is an important figure in any assessment of these technologies. In so far as it is part of policy in all European health services at the moment to keep the sick out of hospitals wherever possible, the GP's role will increase. GP's may be cost effective, but they are generally accepted as hard pressed in a job where time spent with the patient is the main business ... Those new technologies which, in a nutshell, make life easier for the GP are going to be adopted. Technologies which require special training and time, or special staff within a medical practice, may be adopted over a long time and will be used more rapidly in hospitals. (Farrands, 1984)

A similar point is made by Wyke in her review of the problems facing the pharmaceutical industry.

> The new-found concern for medical good housekeeping will also promote the use of more cost-effective drugs. A new product now has possibilities if its manufacturer can show that it gets people out of hospital and back to work faster than the competition. TPA and Eminase fall into this category. Many expect to see more drugs that parade therapeutic economy as their main virtue. (Wyke, 1987)

Similarly many analysts expect to see dramatic growth in the market for new monoclonal-antibody based diagnostic kits, partly because they do not face the same regulatory obstacles that exist for new therapeutic agents that will be taken into the human body. Self-diagnostics is a new growth area, with several new products already on the market in Europe. However their psychological impact on consumers and logistical implications for hospitals have yet to be worked out. Certain specialist areas of medicine might also be seen as developing into market for luxury medical services for those with the money to pay. A strong case can be made for viewing the development of infertility investigations and in vitro fertilisation in this way. (Yoxen, 1986)

It is also clear that human genetics is having an increasingly important effect on medicine after decades of non-interaction. Recent Nobel prizes were awarded for work on the genetics of heart disease. A recent paper in *Nature* posed the question of whether the boundary now lies between infectious and hereditary conditions, if infection can be shown to arise because of inherited susceptibilities, (Diamond 1987). Yet it is not at all clear how such technical knowledge could be put to work in preventive screening programmes. Similarly antenatal diagnosis of a range of genetic disorders continues to expand, driven by medical interest, client desperation and cost-benefit calculations. The impact of such procedures and the difficult decisions they require, on people's self-image, on the full experience of being pregnant or awaiting a birth, against the background of a changing health care system and changing social relations of work, has scarcely begun to be explored. The medical literature on patients' attitudes to and acceptance of new procedures, such as chorionic villus sampling, bypasses these rather more difficult questions. This area is explored in more detail in Nadine Fresco's article in this volume.

## Participation in technology planning and assessment

Even though the latest phase in the evolution of biotechn ology has only just begun, it has nonetheless reached a condition of political visibility normally reached only when technologies are more mature. The fact that such widespread and often heated discussion has gone on so far upstream of major investment in production might be said in itself to be another form of impact on society. Thus recent years have seen very interesting and complex political events, of which the often mentioned, but scarcely analysed moratorium on recombinant DNA research is perhaps the most striking. (Krimsky, 1982) Other public debates and some governmental reviews have also been of interest, because they have addressed directly the issue of non-expert public participation. The consensus conference on biotechnology in Denmark and the Enquete Kommission in the Federal Republic of Germany are two recent examples. It is very important to consider whether such discussions have served to consolidate a tradition of more participatory assessment and control of new technologies.

# PART TWO: HEALTH CARE

# 4 Future Pharmaceutical Markets and Human Health Care

Peter Daly

## BIOTECHNOLOGY AND DRUG DEVELOPMENT

Modern biotechnology results from the discovery of two important new technologies in the 1970s, recombinant DNA (genetic engineering) and monoclonal antibody technology. While the latter technology has had its greatest initial impact on diagnostics rather than therapeutics, recombinant DNA has been used by the industry to produce some novel products such as human insulin, human growth hormone, alpha-interferon, tissue plasminogen activator (tPA) and interleukin-2 (IL-2). Of these the first three are already on the market with marketing approval for tPA expected in the US within one year, although it has already been approved and is on sale in some European markets.

The first generation of biotechnology based drugs are therefore natural proteins which have been produced in large quantities through use of recombinant DNA technology. The cloning of biologically active proteins and their testing for clinical efficiency is incontrast with the traditional method of drug development which is to screen compounds in relevant animal models. Interesting leads are identified and are manipulated by the medicinal chemist and are then evaluated in order to identify analogues with improved potency, duration of action and/or reduced toxicity.

Biotechnology will result in a range of new protein drugs such as IL-2, tPA, superoxide dismutase, erythropoeitin. A number of these will have major market potential. However, it would be mistaken to conclude that the impact of biotechnology in pharmaceuticals is to produce an ever increasing number of protein based drugs. Second and third generation biotechnology drugs will instead involve a synergy between biotechnology and pharmacology and some of the first generation of biologically active proteins will be replaced by better second or third generation products.

Recombinant DNA technology and hybridoma technology will be used to study biologically active proteins and to verify the importance of the proteins in pathology. Recombinant proteins can be used in receptor binding assays to screen for simpler compounds that block the protein's activity. Subsequently, methods will be developed to antagonise the protein leading to the development of a new chemical based drug. Molecular modelling will also be used in combination with these technologies. This technology involves the visual display of macromolecular structures on a computer graphics terminal, the exact measurement of intra- and intermolecular distances and the visualisation of charge distribution and liphophilicity. Molecular modelling will be used to design drugs (small chemical molecules) which fit into receptor binding sites for the biologically active polypeptides which have been identified previously through use of recombinant DNA technology.

In the longer term it should also be possible to design drugs which interact directly with DNA to activate or deactivate genes. For instance oncogene activation is believed to be responsible for the development of many tumours. If molecules could be developed which bond to the oncogene products or regulated oncogene transcription then it would be possible to prevent cells from becoming malignant.

## NEW PRODUCTS BY THERAPEUTIC CLASS

### 1. Cardiovascular

The worldwide cardiovascular market is currently worth $13 billion annually and is forecast to increase to $23 billion by 1991 (Fleming, 1987a) The greatest growth in this market will occur in calcium antagonists which will grow from $1.6 billion in 1986 to $4.8 billion in 1991. ACE inhibitors currently valued at $900 million will increase four-fold by 1991.

Two classes of new drugs will have a major impact in the cardiovascular field. These will be hypolipidaemics which reduce serum cholesterol and fibrinolytics which dissolve blood clots. The leading hypolipidaemic drugs will be lovastatin and velastatin produced by Merck & Co and epstatin from Squibb. These inhibit the enzyme HMG CoA reductase, which controls a key step in cholesterol synthesis, and so lowers the cholesterol level. Lovastatin is currently awaiting registration in the US and elsewhere and is expected to be the first drug in this new class of hypolipidaemics to reach the market. It acts, as does its follow-up product, velastatin, to reduce total plasma cholesterol and in particular LDL-cholesterol. The market for this and similar products is expected to be large and could reach $1.3 billion by 1991 (Fleming, 1987)

Much media attention has recently been given to the new class of fibrinolytics and in particular to tissue plasminogen activator (tPA) produced by Genentech and other firms through genetic engineering techniques. Fibrinolytics such as tPA and its main competitor, Eminase (Beecham), will be given to patients within minutes or a few hours after myocardial infarcts and will result in clot lysis. Tissue plasminogen activator is the first major new genetically engineered drug to emerge and a market of around $1billion by the early 1990s has been predicted for it. However, there is some uncertainty over the patent situation and a wide ranging tPA patent by Genentech has been successfully challenged recently in a British court.

Eminase has a competitive advantage over tPA in dosing/administration (bolus iv could be administered by a paramedic), patent protection and price, while tPA has a faster onset of action (20 versus 40 minutes) and possibly greater safety.

The enzyme superoxide dismutase may also be used as an adjunct to fibrinolytic therapy. Recent clinical data shows that damage to heart muscle is caused by the action of free radicals generated upon re-perfusion. Use of superoxide dismutase which acts as a free radical scavenger may prevent this problem.

Another cardiovascular product where biotechnology is being applied is atrial natriuretic peptide (ANP), a heart hormone which is involved in controlling blood pressure and volume. This has potential use in hypertension and congestive heart failure. One company which is developing this is California Biotechnology which has cloned the gene for the ANP receptor and is studying receptor structure in order to design synthetic compounds with ANP-like activity but which can be delivered in oral form. (California Biotechnology, 1986). This is an approach to drup development which will become increasingly important as biotechnology advances. California Biotechnology is also developing diagnostic tests which will indicate susceptibility to atherosclerosis and hypertension based on genetic data.

Biotechnology techniques including recombinant DNA will be used to study a wide range of systems associated with heart function and disease including LDL receptors, cholesterol transport and biosynthesis, and factors involved in blood pressure regulation. The increased information on these systems will permit development of new and more effective therapies for cardiovascular disease.

## 2. Antibiotics

Antibiotics traditionally include a range of product classes including penicillins, cephalosporins, macrolides etc. Three classes of antibiotics are expected to dominate new product activity in infectious diseases. These are arbapenems, monobactams and quinolones. Quinolones in particular such as Merck and Co's Noroxin and Warner-Lambert's Comprecin are likely to become important anti-infective products over the next few years. Quinolones have a broad spectrum of action, an oral route of administration and show continuing bactericidal activity at low plasma levels. This latter property suggests that missing a dose may not be a serious event in therapy of infectious diseases with quinolones.

Biotechnology will be applied in antibiotics to optimize and improve existing processes and also to generate novel antibiotics which could not be made by traditional organic chemistry techniques.

One example of possible biotechnology applications to processing is in the production of cephalosporins. A key stage in the manufacture of cephalosporins is the biosynthesis of cephalosporin C. The gene coding for one of the biosynthetic enzymes in cephalosporin C biosynthesis has been cloned and research is continuing to clone the gene responsible for the other enzymes. When this has been achieved it may be possible to modify the structure of these enzymes (through site-directed mutagenesis) and so produce cephalosporin C more efficiently (Balz et al., 1986).

Recombinant DNA technology is also being applied to the production of novel antibiotics. Researchers at Eli Lilly headquarters in Indianapolis are using recombinant DNA to develop novel macrolide antibiotics related to erythromysin. This latter antibiotic is produced by the organism *Streptomyces erythraeus*.

The strategy employed by the Eli Lilly researchers is as follows. They have cloned a fragment of DNA (35 Kb) which encodes the entire biosynthetic pathway of erythromysin. They are now attempting to clone the biosynthetic pathways for other macrolide antibiotics produced by other *Streptomyces* species. When this is achieved it will allow the *in vivo* recombination of biosynthetic pathways and lead to the production of novel hybrid antibiotics.

Recombinant DNA research concerned with the manipulation of biosynthetic pathways is at an early stage of development. Consequently, we cannot expect biotechnology to have much effect on antibiotics in the short term. However, in the long term (10 years) it will have a significant impact.

### 3. Psychotherapeutics

Psychotherapeutic drugs cover a wide range of chemical entities and indications including anxiolytics, antidepressants, drugs used in the treatment of schizophrenia and Parkinson's disease and anticonvulsants. Also included here are central nervous system (CNS) drugs for brain injury.

At a symposium on Preclinical Strategies in Psychopharmacology, held in Paris in January 1987, speakers discussed the fact that the classical major diagnostic groupings in psychiatry are breaking down, as it becomes realised that different aspects of apparently the same illness respond to different types of drugs and the strategy of using different agents at various stages of a disease gains acceptance among clinicians. (Scrip, March 20 1987)

Neurobiology is undergoing a major revolution today which is changing our concepts of the brain and central nervous system; consequently psychotherapeutic and CNS drugs are in a state of flux. One of the most important CNS diseases in view of the "greying" of western populations is Alzheimer's disease. This disease which is characterized by specific cholinergic degeneration is the focus of much research actvitiy by pharmaceutical companies. In the US about $40 billion is spent annually on caring for Alzheimer's patients. One approach investigated by various companies is the use of neurotrophic factors such as nerve growth factor and fibroblast growth factor. Another drug, THA (tetrahydroaminoacridine), may be temporarily useful in treating the condition and is receiving much attention. A bill that would provide $6 million for fast-track clinical trials with THA has been introduced in the US Senate. However a drug which can stop and/or reverse the disease has yet to be developed. Molecular biology has a very important role to play here in terms of identifying aberrant gene products and their role in neuronal degeneration.

A more immediate area of development is in the area of serotonin antagonists and uptake inhibitors. In the UK both Glaxo and Beecham have 5HT3 antagonists in clinical trials for such indications as emesis, schizophrenia and anxiety. The emesis use will be for those undergoing cancer chemotherapy. Eli Lilly has a serotonin uptake inhibitor, Prozac, which is not only useful in depression but also reduces appetite and obesity.

4. Antivirals

The market for antiviral drugs is growing rapidly and is expected to increase from $150 million in 1985 to $680 million in 1990 (Fleming, 1987b). There are a range of important viral infections for which such drugs will be used including genital herpes, cytomegalovirus, influenza, hepatitis and, most important of all, AIDS. None of the antiviral drugs on the market or in clinical trial can completely cure viral infections, in the sense of elimination of the virus from the body but they block viral replication, so controlling or reversing the disease.

The most serious viral disease today is, of course, AIDS. There are a number of drugs which are being used against AIDS. Of these the one which has received the most attention is AZT from Wellcome. a recent study of AZT showed that of 19 individuals tested of over a 6 week period the number of helper T-lymphocytes increased in 15 of the 19 individuals. Two individuals had spontaneous remission of fungal infections and the virus could not be detected in individuals who received the highest doses. Another compound related to AZT is dideoxycytidine (DDT) which has been licensed to Roche but is the subject of patent litigation by Wellcome. This has similar effects to AZT in relation to weight gain and T4 cell rise but without the severe bone marrow suppressive effects of AZT.

An entirely different type of drug, Ampligen, has recently been shown to have significant effect on AIDS. This drug is a mismatched double stranded RNA which has been developed by the US Company HEM and licensed to Du Pont. A study on ten patients reported in the Lancet showed that 4 AIDS patients has a transient 2.5 fold increase in T4 cells at four weeks of therapy together with rapid shrinkage of lymph nodes. In 9 patients, who were seropositive for HIV, RNA levels in peripheral blood mononucleocytes became undetectable between 10-40 days of starting therapy.

A different approach to the sole use of viral blocking agents in AIDS is the use of immunomodulating drugs including lymphokines, such as alpha-interferon and interleukin-2, in combination with drugs such as Wellcome's Retrovir and Ampligen. AIDS therapy in the future is likely to consist of 'cocktails' composed of viral replication- blockers and immunomodulators and Wellcome is just starting to experiment with a mixture of its alpha-interferon in combination with Retrovir. A synergistic effect has also been observed through use of Ampligen and Retrovir in combination.

Lymphokines show encouraging results against a number of other viral diseases. Alpha-interferon in the form of a nasal spray can prevent the common cold. However over-use can cause symptoms which mimic those of a cold and it is also very expensive. Hence its potential success in this area is uncertain. Alpha-interferon is marketed at present for treatment of leukaemia and other cancers and genital warts. It also has considerable potential for the treatment of chronic hepatitis. Other immune modulators such as interleukin-2 (IL-2) are also being investigated for possible antiviral use.

5. Anticancer drugs

The anticancer drug market is entering a major period of change brought about by the effects of biotechnology. Since 1982 an extraordinary amount of progress has been made in our understanding of the molecular basis of cancer through the discovery of oncogenes. This period has also seen the introduction by pharmaceutical companies of alpha-interferon products for leukaemias and Karposi's sarcoma including Schering-Plough's Intron A, and Hoffman La-Roche's Roferon. Both of these products consist of a single type of alpha-interferon produced by genetic engineering while Wellcome's Wellferon consists of a mixture of alpha-interferons produced by cell culture.

A range of other lymphokines and related immunomodulators is also under development by pharmaceutical and biotechnology companies. The US company Cetus has developed interleukin-2 analogues and the company is now seeking European patents through the European Patent Office in addition to its US patents. The use of IL-2 in association with lymphokine activated killer cells (LAK) in a US National Cancer Institute trial received much publicity in 1986. Interleukin-2 therapy is being tested on patients with a wide range of tumours including metastatic melanoma, refractory lymphoma, lung adenocarcinoma, ovarian cancer and others. The drug can have severe side-effects including cardiac toxicity, fluid retention, fever and nausea.

Another recombinant immunomodulator in clinical trials is tumour necrosis factor (TNF) which is being developed by a number of companies including Genentech, Cetus, Dainippon and Hayashibara.

A completely different approach to cancer therapy is the use of monoclonal antibodies to target chemotherapeutic drugs, toxins or radionuclides to tumour cells. Pharmaceutical and biotechnology companies such as Centocor, NeoRx, Eli-Lilly and Xoma are developing products in this area. Centocor has developed a number of *in vitro* diagnostic tests for various cancers using monoclonal antibodies and is now using some of these monoclonals for therapeutic applications. Eli Lilly is carrying out research on the KS 1/4 antibody linked to cytotoxic drugs for use against lung, breast, prostate and pancreatic cancers. Xoma Corporation is using immunotoxins containing the powerful cell toxin ricin in trials against melanoma and colorectal cancer. NeoRx is developing radionuclide, toxin and drug monoclonal conjugates. It is using an iodine-131 conjugate and a *Pseudomonas* exotoxin conjugate has entered trials against adult T-cell leukaemia and a mechanism has been discovered to increase dramatically the potency of drugs that are poorly cytotoxic in their unconjugated form.

Research is also underway in many companies to develop new non-biotech drugs or to develop less toxic analogues of existing chemotherapeutic drugs. Research here may be oriented towards increasing anti-tumour activity or decreasing toxicity. New classes of anti-neoplastic agents are being identified such as the alkylphosphocholines which act on tumour cell membranes and which are being developed for skin mammary carcinoma. Biotechnology products such as immunotoxins and lymphokines will not replace cytotoxic drugs. Instead, various 'cocktails' of cytotoxic drugs and lymphokines will be used and conventional anticancer drugs will also be required as conjugated for monoclonal based products.

## HEALTH CARE POLICIES AND COST CONTAINMENT

The containment of healthcare costs is an important issue to governments worldwide and cost containment measures are one of the most important factors in the environment in which the pharmaceutical industry operates. Different types of cost containment measures have been introduced in various European countries reflecting their different financing of healthcare costs. In the UK, where healthcare is provided almost entirely by government and financed by tax revenue, a limited list has been introduced, prescription charges raised and smaller hospitals have been closed. In Germany certain types of drugs are also no longer reimbursed, and since the creation in 1977 of the Concerted Action Committee for Health Affairs, targets have been set each year for the growth of health care expenditures. (Eurocare, 1988) The guidelines of the Committee are not themselves binding but excessive expenditure by one sector may lead to tighter monitoring of that sector's expenditure. Consequently, the health professions and interest groups follow the targets set by the Committee, if they want to keep their independence.

In the US reimbursement is now based on diagnostic related groups (DRGs) which establish a maximum amount payable for a particular illness. There has been a rapid increase in healthcare delivery through Health Maintenance Organizations (HMOs). These deliver a defined set of medical benefits to a specific population in return for a fixed annual payment per head. If the HMO can provide medical care for less than the average capitation fee it makes a profit; if it exceeds that fee it loses money. By 1995, half the US population will receive healthcare from managed care systems such as HMOs and preferred provider organizations.

Cost containment measures are having and will continue to have significant effects on drug use and drug development. Proof of comparative cost effectiveness will be the main consideration determining registration, reimbursement and addition to drug lists, including State, hospital and insurer-approved drug lists. There will be little room for 'me-too' drugs and companies will be under greater pressure to come up with innovative products. Cost effectiveness will cover a range of criteria. Drugs will need to demonstrate cost-effectiveness in reducing hospitalization periods and in permitting early possibility of ambulatory care. Cost-effectiveness in relation to lengthening disease relapse period and increasing patient compliance will also be relevant. An important issue also will be the contribution by a particular drug to a reduction in hospital labour costs eg. through prolonged therapeutic activity, which reduces the need for frequent administration.

These forces should in general work in favour of biotechnology-based drugs. These drugs are by definition innovative and in many instances they will represent the introduction of greatly improved therapy for various conditions. Price however will be an important determinant of their acceptability. Inital prices may be high (eg. tPA from Genentech will cost about $1,000 - 2,000 per procedure) but these prices will drop sharply in some instances, as competition develops and it is certain that intense competition will exist in some markets. For instance there are about 30 companies worldwide involved in R & D on tPA and this does not include other fibrinolytics which will compete with it. A similar situation exists in relation to interleukin-2. This price competition may be further increased by the emerging patent situation for recombinant proteins, which is very uncertain and which shows signs of moving towards a situation in which genetically engineered proteins would either not be patentable or impossible to defend. Important 'breakthrough' drugs developed through biotechnology will rapidly establish themselves as the preferred choice of therapy among physicians and will therefore be integrated rapidly into reimbursement systems.

## IMPACT ON HEALTH CARE

Therapeutic advances over the next ten years will undoubtedly have a major effect on improving healthcare in the developed countries. How health care is delivered will depend however on the regulatory factors described above and on social changes. In this section some ideas are presented on how advances in biotechnology and pharmaceuticals will affect health care up to 1997.

The main factors within which the industry will operate will be as follows:

- cost containment measures will continue to dominate the environment within which the pharmaceutical industry operates

- the proportion of elderly people in the population will increase significantly

- there will be increased emphasis on self-medication, self-diagnosis and prevention

- a greater proportion of surgical procedures performed on an outpatient basis

Within this environment the chief therapeutic benefit will be:

- a greatly improved cure rate for cancer

- an effective therapy for AIDS

- major advances in cardiology, rheumatology and CNS diseases

- effective therapy for Alzheimer's disease

## EXPECTED THERAPEUTIC ADVANCES

### 1. Cancer

Cancer prevention will assume much greater importance in 1998 than today with emphasis on diet and the elimination of smoking and environmental pollution. The early detection of cancer through rapid diagnostic tests based on monoclonal antibody technology will also assume greater importance.

In relation to therapy, the overall cure rate will increase from about 50 per cent today to 70-80 per cent by 1997. In the longer term with our increased ability to interfere with oncogene activation, this cure rate will rise further. The greatest degree of improvement is likely to occur in leukaemia with solid tumours presenting greater difficulty. The delivery of cytotoxic drugs by monoclonal antibody technology is likely to become more routine and cytotoxic drugs will also be combined with lymphokines. The improved survival rate from cancer will be a further factor tending to increase the proportion of old people in the population.

2. Aids

Many present day forecasts on AIDS envisage the disease advancing inexorably into the 21st century. These forecasts do not take into account two very important facts: firstly our rapidly increasing capabilities in molecular biology, immunology and virology and secondly the encouraging clinical trial results from existing antiviral drugs. Since the AIDS virus was identified, its entire genome has been sequenced, monoclonal antibodies have been raised against viral antigens and much has been learned about its interaction with the immune system. Had the AIDS crisis occurred in the 1960s instead of the 1980s none of this would have been possible. The prospects for using human monoclonal antibodies therapeutically rather than just in diagnostics are good and our understanding of viral replication and relationship to the T-cells in increasing at a very rapid rate. Another important consideration is the progress being made with existing antiviral drugs is showing considerable potential for blocking viral replication and restoring immune competence and this is very encouraging for the development of more powerful second or third generation products. Consequently, it is likely that AIDS will be greatly diminished as a public health issue by 1997 and effective therapy will be generally available in Western countries.

3. Cardiovascular disease

As in the case of cancer, prevention will do more than therapy to reduce the burden of disease. Prevention will however involve an increased use of certain drug categories including lipid lowering agents, plaque dissolving drugs and antihypertensives. Fibrinolytics will be used extensively by the late 1980s and by the early 1990s drugs which minimise the damage to heart muscle after a heart attack will be introduced. These drugs will make significant contributions to mortality arising from heart attacks.

Disease susceptibility diagnostics based on RFLP (DNA probe) technology could make a major contribution to the prevention of cardiovascular disease if they were widely used in society. Such tests carried out at a relatively early age might then allow appropriate modifications in diet and lifestyle. However, as in other areas where DNA probe technology is involved, there could be ethical issues associated with use of such tests and the identification of susceptible individuals and groups.

4. Inflammatory diseases

The present drugs used in rheumatology are the non-steroidal anti-inflammatories (NSAIs). These treat only the symptoms and do not affect the underlying disease process and also have serious side-effects such as the induction of ulcers. Much research is currently underway into developing disease modifying agents and such products will probably be introduced for arthritis in the early 1990s. This development combined with improved analgesics should result in a great improvement and a lessening of the morbidity associated with it. Such drugs will contribute to a greater degree of mobility for older patients.

Asthma and other respiratory conditions also have an important inflammatory component and the development of leukotriene receptor antagonists will also improve therapy for these conditions.

## HEALTHCARE AND THE ELDERLY

By 1997 there will be a much larger proportion of elderly people in the Western population. These increased numbers will translate into increased political influence which will effect public health policy in such areas as insurance for long term care and drug testing in the elderly.

The elderly will present a large market opportunity for pharmaceutical companies and diseases of the elderly CNS diseases especially will receive greater attention in R & D programmes. One disease which is of major proportions is Alzheimer's disease. This disease is especially stressful for relatives who must tend to the patient over an extended period as he/she degenerates mentally and physically. By 1997 drugs should be available which will slow or prevent neuronal loss. The disease is estimated to affect 2.5 million people in the US alone. Drugs which alleviate or cure the condition will make a major contribution to improving the quality of life of the elderly patients and their relatives.

## SELF MEDICATION AND DIAGNOSIS

Cost sharing in the treatment of minor ailments is now established in the healthcare systems of European countries. The creation of positive and negative lists of prescribable drugs in particular is leading to an increase in direct purchase of medicinal products by the consumer who may be advised by a pharmacist. At present the main uses of over the counter (OTC) drugs are for pain, colds, coughs and digestive disorders. An increasing number of medicines will become OTC over the next few years and in 1989/90 Smith Kline plans to introduce an OTC version of its leading anti-ulcer drug cimetideine. Many products are unsuitable for self-medication but some in the dermatological and cold remedy area could be marketed as OTC products. For instance, proteins with anti-ageing effects could be marketed as cosmetics and alpha-interferon could be sold as a nasal spray for colds.

In self-diagnostics, products are already on the market for such uses as pregnancy, fertility and therapeutic drug monitoring. This market sector is expected to expand rapidly and major growth areas will include therapeutic drug monitoring, infectious diseases, sexually transmitted diseases and reproductive-related tests. Cancer detection in home-based kits is technically feasible but other factors suggest that this will not be widely applied. Home tests for cancer, conducted and interpreted by the user ouside the control of an established programme, carry the risk of psychological damage, if professional guidance is not obtained. Consequently, self-diagnosis for this indication is unlikely to be developed.

In summary, individuals will assume more control over their own health, through prevention, self-medication and diagnosis and the population will be much better informed on medical matters. When individuals do fall ill, a greater range of more effective therapeutics will be available.

# 5 Social Dimensions of Biotechnology: The Case of Vaccines

David Banta

## INTRODUCTION

Most vaccines at present are preparations of killed or attenuated (weakened), and thus nonvirulent, micro-organisms that are injected into an individual in advance of the person having a particular disease, with the aim of producing immunity to the disease. The nonvirulent micro-organisms injected carry complicated molecules, called antigens, on their outer coat. Lymphocytes of the recipient's immune system recognise these antigens, which may be proteins, carbohydrates or lipids, as invaders and respond by producing antibodies that bind to the antigens and identify them for destruction by other components of the immune system. In particular, when bound to their target, they cause it to become attached to a receptor present on the surface of white blood cells, which then engulf and destroy the enemy thus branded. Antibodies also serve to alert a soluble killing system carried by the bloodstream and known as complement. Antibodies circulate in the body for a certain time, protecting a person against later exposure to the disease-causing organism. A person then is partially or totally 'immune' to the disease.

Vaccines are by far the most effective and inexpensive of available methods for prevention and control of communicable diseases (Chin, 1986). Smallpox has been eradicated from the world by an international vaccine campaign (Behbehani, 1983), and world elimination of measles and rabies are currently being viewed as additional worthy targets (Beran and Crowley, 1983; Foege, 1984). Measles, poliomyelitis, diphtheria, pertussis (whooping cough), and other diseases have already been brought under control by widespread vaccine use. The reduction in the number and size of epidemics of infectious diseases in this century is largely due to vaccination (Perkins, 1973).

The new biotechnology has presented a number of possible future prospects, including the development of a number of new vaccines, improvement of existing vaccines and, if possible, production of more inexpensive vaccines (Bialy, 1985). New approaches will be of particular value where organisms cannot be cultivated by conventional methods, such as in the case of the hepatitis B vaccine, or where highly pathogenic and contagious organisms are involved so that their cultivation carries a risk to human health. In addition, improvements in existing vaccines can result in fewer side-effects, as in the case of pertussis. New vaccines made by traditional methods will probably become available in the near future.

Many new vaccines, such as an improved cholera or a new malaria vaccine, will be most useful in tropical developing countries and may have large effects on the health status of populations of those countries. Of course, their ultimate use in these countries will be determined by a number of factors, not only technological success. In industrialized countries, too, a number of infectious diseases are still prevalent, or new infectious diseases, such as AIDS, are emerging. Many of these can possibly be prevented by new vaccines (Ruidenberg, 1986). A partial list of diseases of importance include Bordetella pertussis (improved vaccine), cytomegalovirus, haemphilus influenzae, herpes simplex viruses 1 and 2, influenza (improved vaccine), and AIDS.

## THE SIGNIFICANCE OF VACCINES

Vaccines have now been in use for more than 200 years. The idea of immunity to infection is an old one. For example, it has long been known that a person who had one attack of smallpox would not have another (Sigerist 1970, pp 285-288). In India, children were wrapped in clothes that had been worn by those with smallpox to induce a (hopefully) mild form of the disease and subsequent immunity. Edward Jenner believed that infection with the benign cowpox virus protected against smallpox. He tested this theory in 1796 by vaccinating a boy with cowpox organisms and demonstrating his subsequent immunity to smallpox. Pasteur was perhaps the greatest innovator in vaccines. He weakened organisms to develop vaccines for chicken cholera, anthrax, and swine fever and then for rabies. The techniques developed by Pasteur made possible immunization against such common diseases as tetanus (beginning in 1907), pertussis (1914), typhoid (1914) and diphtheria (1926). The isolation of viruses in 1935 and the ability to grow them in culture made possible the development of vaccines for influenza (1945), poliomyelitis (1955), measles (1962) and rubella (German measles) (1969) (OTA, 1979, pp 155-156). (Dates given in brackets are dates of the introduction of the vaccine in the United States).

The impact of vaccines may be illustrated by the case of smallpox, which was historically the most important communicable disease (Henderson, 1986). Smallpox was endemic (occurring in non-epidemic form) throughout most of the world. When it was introduced to America during the Spanish Conquest, its impact was devastating; in Mexico alone, more than 3.5 million people died. In Europe and North America, smallpox persisted until World War II. The last case of the now extinct disease was seen in Somalia in 1977. The total cost of international assistance to developing countries in the worldwide campaign against smallpox was US$112 million, an amazingly small figure. (Henderson, 1986)

Fudenberg carried out one of the early analyses of the costs and benefits of vaccines, using polio vaccine. (Fudenberg, 1983) He estimated that 154,000 cases of polio were prevented in the United States between 1955 and 1961. Of these, he estimated that 12,500 would have died and 14,300 would have experienced complete disability. The medical care costs for this group of people would have been an estimated US$327 million and the loss of earned income over their lifetime was estimated at US$6.4 billion. He estimated the cost of avoidance - the cost the of vaccine, the cost of administration, the cost of physicians and nurses - to total US$611 million.

This story has been repeated again and again. It is this striking success of vaccines in the past that has raised hopes that the new techniques of biotechnology can lead to further vaccines to protect human health. Vaccines are technologies that are more involved with public policy decisions than many others. Traditionally, governments have made special efforts to develop vaccines. In the case of some, governments also make concerted attempts to encourage, or even require, their use. As biological products, they are regulated to minimize risks and assure benefits; all industrialized countries have mechanisms to examine their efficacy and safety before they are widely used.

## COST EFFECTIVENESS OF VACCINES

Some of the earliest cost-effectiveness analyses done in the health area concerned vaccines, and this continues to be an active area of analysis. The main reason, most likely, is that vaccines are generally paid for by public tax funds. Setting priorites among competing investments of tax funds is obviously difficult, and cost-effectiveness analysis can help indicate whether the investment is wise or not.

Ten vaccines have been examined for their cost-effectiveness: measles (7 studies); pneumococcal vaccine for pneumonia (4 studies); mumps (4 studies); rubella (4 studies); influenza (4 studies; polio (2 studies); pertussis (2 studies); and hepatitis B (1 study).[1] All are financial successes, in the sense that benefits, as valued, exceed costs for some target groups and in some situations. Still, almost all of the studies neglect some benefits, including quality of life and the saving in time and effort for parents who must care for a sick child (Weisbrod and Huston, 1983). In short, vaccines are even more cost- effective than indicated by the literature.

As is apparent, the majority of the analyses have concerned childhood vaccines. Most of these vaccines are money-saving in use because of the health care costs associated with the disease. Vaccines against childhood diseases are generally used intensively in industrialized countries, and vaccination rates are often 90 percent and above.

An interesting example of a vaccine that has never been widely used is influenza vaccine. This has been the subject of four studies. All analyses find it to be cost-effective, but benefits exceed costs only for high risk persons (Klarman and Guzick, 1976). The Office of Technology Assessment did a very comprehensive analysis of this vaccine and found it to be cost-effective (OTA, 1981). The average cost of increasing healthy life expectancy ranged from about US$2,000 for a year of quality-adjusted life in the worst case to money saving in persons over the age of 44 with the best set of assumptions.

[1] References to these studies are available from the author.

The analyses of pueumococcal vaccine find the vaccine cost-saving only when used in high-risk groups, primarily the elderly. The Willems et. al. study, published in longer form by the US Government Printing Office, is interesting because it led the US Congress to cover the vaccine under the Medicare program. (Willems et al., 1980; OTA,1981). The cost of a quality-adjusted life year was found to be US$1,000 for those over the age of 65.

Finally, Mulley et al. carried out an analysis of the costs and benefits of the hepatitis B vaccine marketed beginning in 1982. (Mulley et al., 1982) The problem with this vaccine, from the standpoint of cost-effectiveness analysis, is that it is very expensive. The analysis found that the most effective strategy for vaccinating homosexual men, a high risk group, was to screen for hepatitis antibodies first, and then give the vaccine only to those with no antibodies. This alternative cost about US$66 per person, compared with about US$100 per person for vaccinating all. However, for those countries with a high prevalence of hepatitis B and associated liver cancer, the calcuations may be quite different. For a medium risk population, surgical residents of hospitals, vaccinating everyone was the lowest cost option, costing about US$104 per person. Wider use of this vaccine seems to depend on lowering its cost. (Weisbrod and Huston, 1983)

Despite the evidence on these three vaccines addressed to diseases in adults, they are not widely used. The next section discusses this issue in more detail.

## FACTORS IN UTILIZATION OF VACCINES

Whether a vaccine is cost-effective or not is not the only factor in determining whether it will be used. Even if social policy is to encourage a certain vaccine, people, including health care providers, must be willing to accept it.

Factors affecting utilization of a vaccine include the following: 1) vaccine availability; 2) the effects of statutory interventions, such as school immunization requirements; 3) characteristics of the target population, including its access to health care providers; 4) characteristics of the vaccine that may affect the provider, including route of administration, storage conditions, cost of delivery and special procedures required; 5) characterisitics of the vaccine that may affect patient acceptance (number of doses, route of administration, adverse reactions, cost; and 6) target population attitudes toward the vaccine and the disease. (Institute of Medicine, 1985)

Factors that physicians may consider in assessing a patient's need for a particular vaccine include: 1) the likelihood of the patient's being exposed to the particular organism; 2) the patient's vulnerability to the disease once exposed to the organism; and 3) the extent to which contracting the disease will cause mortality or morbidity or disrupt the patient's life. (OTA, 1979, p. 181) A particular problem is the data base that physicians use to make these judgements. For old vaccines and well-controlled diseases, data are generally available and physicians are familiar with evidence from their clinical training. For new vaccines, however, physicians rely on traditional sources of information, including peers and professional literature. Vaccines are not generally actively promoted by industry. Government may not take an active role in education. Perhaps these factors explain why physician acceptance of a new vaccine may be quite low (Pantell and Stewart, 1979). Recently, awareness of complications of vaccines has been growing, and some countries, such as the United States, have faced the problem of compensation for vaccine-related injury. Publicity surrounding this problem may have discouraged physicians from prescribing new vaccines (OTA, 1980, p. 181).

Despite recommendations of public health officials or physicians, vaccine use (beyond those required by law) depends on acceptance by lay people. Acceptance of a vaccine depends on such factors as the perception of the danger of the particular disease, beliefs regarding the efficacy and safety of the vaccine, and the convenience of being vaccinated (Ibid, p. 175).

Negative publicity around a vaccine may have serious consequences. An example is the rapid fall in coverage with pertussis vaccine, especially in the United Kingdom and the Federal Republic of Germany after dramatic publicity concerning the side effects of this vaccine. In Sweden and Japan, pertussis vaccination had been officially stopped. In these countries, pertussis disease (whooping cough) reoccurred, producing far more damage and even death than the side effects of the vaccine. In the Netherlands, the authorities have warned against the reoccurence of the disease, and as a consequence coverage with pertussis vaccine did not drop below 90 percent.

The 1976 experience in the United States with the emergency vaccination program for swine influenza, with its serious side effects, eroded public confidence in new vaccines (Koplan and Kane, 1986). The fact that hepatitis B vaccine was made from serum had led to fears that it might be a carrier of AIDS. The release of the recombinant DNA hepatitis B vaccine ended this potential threat. However, can the average member of the public make a realistic assessment of risks? In general, little effort has been made to educate the public concerning the benefits, risks, and costs of vaccines. In the United States the Institute of Medicine has judged provider and consumer acceptance of a wide variety of possible future vaccines, using the categories risk, severity, benefits, and barriers. (Institute of Medicine, 1985) Such information needs to be more widely available.

The new biotechnology could improve acceptance of vaccines in a number of ways. One would be to produce safer vaccines, with minimal adverse reactions. Another would be to produce polytopic vaccines containing immunologically important parts of several organisms, cutting down the number of vaccine administrations. Finally, development of new methods of administration of vaccines, such as nasal sprays and oral vaccines, could play a role.

The effect of the cost of vaccination on vaccine acceptance has not been asssessed. One must assume that the cost of hepatitis B vaccine - in the Netherlands about Dfl. 200 (US$100) per course of 3 injections - affects utilization. The desirability of vaccines from the standpoint of public health has often led to provision of vaccines free of charge.

## SOCIAL POLICY TOWARDS VACCINES

Rates of immunization in the population have varied rather dramatically over time, and are related to a number of factors, some discussed above. Since vaccines are preventive measures whose benefits occur sometime in the future and are not spectacular, they may not be sought as actively as other health care technologies, whose benefits may be more immediate.

The importance of social policy may be illustrated by the case of measles vaccine in the United States. When the Federal government has provided money for measles vaccine, the number of immunizations has risen, and the number of cases of measles has fallen. When the Federal government has not provided money, the number of immunizations has fallen and the number of cases of measles has risen.
(OTA, 1979, p. 183).

Because of this fact, and because high rates of immunization are necessary to prevent epidemics, governments have provided vaccines as part of their public health programs. Governments often require certain immunizations among their population, especially for children entering school. A typical requirement is that a child would be required to be immunized against tetanus, diphtheria, pertussis, measles, rubella and mumps. The vaccines are generally provided free or at low cost through public health programs.

For vaccines against diseases prevalent in the adult population, the situation is not so positive. The two main vaccines in question are influenza vaccine and pneumococcal vaccine (mainly to prevent bacterial pneumonia). These vaccines are provided sporadically through public health programs, and few countries have achieved high rates of immunization against these problems (Koplan and Kane, 1986). In the United States, where pueumococcal vaccine is provided to all elderly people through the Medicare program, immunization rates nonetheless continue at a low level. In the future, new vaccines will encounter similar problems of acceptance by professionals and the public alike. New vaccines against childhood diseases are much more likely to be accepted than others, since the child is under active physician surveillance and the mother makes most health care decisions. In general, vaccines will not the given to the entire population of a certain age, but only to those at risk. The problem may be illustrated by the case of a future vaccine against gonorrhea. The population at risk must not only be identified, but must in some way be convinced to seek and accept the vaccine.

In the future, governments may need to consider much more active attempts to 'educate' the public concerning vaccines. In short, if the government of a country is convinced, by cost-effectiveness analyses or other studies, that a vaccine should be widely used, will it seek to influence behaviour of the public? If so, how will such attempts be developed and structured?

## CONCLUSIONS

The field of the new biotechnology is very complex, and in some ways is only at a beginning. Although striking advances have already appeared, as work goes on, difficulties are also becoming obvious. Some early - and very optimistic predictions - about future advances have not yet materialized. More fundamental knowledge is clearly needed before certain expected advances can appear. This does not mean that biotechnology does not have great promise for the future. It does mean, though, that the importance of continued support for research in this area must be recognized.

In the past, vaccine development efforts have been generally uncoordinated and fragmented. With the new capabilities, this situation is increasingly dysfunctional. In the United States, the Institute of Medicine has recommended priorities for developmental efforts. Other countries engaged in vaccine development need to examine their policies in similar ways. One particular problem is the general weakness of data systems concerning the incidence, prevalance, and social importance of infectious diseases. In addition, international cooperation and communication are increasingly important.

The acceptance of vaccines by the general public is a critical issue. Maintaining immunization rates against childhood disease above 90 per cent, for example, totally prevents epidemics of childhood diseases that once ravaged the population. The highest priority should be given to maintaining such immunization rates.

Most vaccines in present use are given to all members of the population in a certain age group, usually children or the elderly. In the future, a number of new vaccines will have to be targeted to particular sub-groups in the population to be both effective and cost-effective. Behavioral science research specific to the question of acceptance of vaccines could be of value.

Infectious disease is not a national issue, but an international one, especially in Europe, with its small coutries and its open borders. With better vaccines, improved control and even elimination of some diseases is feasible, but this will be most effective if done internationally.

# 6 Biotechnology in Human Genetics: Developments and Problems

Nadine Fresco

## INTRODUCTION

In the report which he presented to the European Foundation in November 1986 on "The Impact of Biotechnology on Living and Working Conditions", Edward Yoxen wrote: "Biotechnology can be spoken of as a cultural phenomenon. It is not just hardware, organisms and technical workers. It also exists as a vision of what the technology could usher in, a vision rehearsed in countless press articles, company prospectuses, business surveys and government reports." (Yoxen, 1987, p. 13) The fact that biotechnology is a cultural phenomenon is in itself a broadly sufficient explanation why, as with all cultural phenomena, there has been such a copious production of written texts, of varying origins and qualities. But there is an additional reason for this output, which derives from a particular field of application of biotechnology - the human person - and the vision of the future in which this application is embedded. Very frequently one finds that the endless articles and reports devoted to these questions offer by way of introduction, or more frequently by way of conclusion, general humanist observations on the collective wisdom which will prevent the emergence of a "best of all possible worlds" populated by Frankenstein's creations, or else alarmist warnings on the imminent danger of such a scenario.

What present and future applications of biotechnology to human beings provide the foundation for these optimistic or pessimistic commentaries? Some of them derive directly, for the moment at least, from the field of medicine, while others do not. DNA analysis has been used for about ten years in the pre natal diagnosis of certain single-factor genetic disorders. A distinction is to be drawn between direct DNA analysis, most often applied to cells obtained from the foetus, and the semi-direct or indirect analysis involved in a genetical study of the family.

Direct analysis makes it possible to identify and analyse specific genes through costly techniques using so-called DNA probes (involving the hybridisation or chemical bonding of a DNA fragment, the nucleotide sequence of which is wholly or partially homologous to that of the gene in question). In essence the entire set of genes of an individual are searched for a small section which matches the specially designed probe. This is already possible with sickle-cell anaemia and certain forms of beta-thalassemia.

Semi-direct analysis, which reveals not the mutant gene itself, but the fact of internal changes to the gene through the displacement of a nearby genetic marker, requires a complete family study to be carried out before the pre natal diagnosis, which makes it particularly cumbersome. But it is often the only possible method, however, when diseases are caused by any one of a number of detailed mutations which are not detectable by direct diagnosis. This is the case with haemophilia B or phenylketonuria. The latter is a metabolic disorder leading to severe mental retardation, now being systematically tracked down at birth in Western countries - the frequency of the disease in France being 1 in 16,000 births - and treated with a very long and restrictive dietary regime.

Indirect analysis concerns cases in which DNA probes cannot be used, because the gene involved has not been identified. This type of analysis also needs an exhaustive study of the family involved - a study which is often very difficult to carry out. It is already effective, or on the point of becoming effective, for Huntington's chorea, a degenerative disease of the nervous system which only appears in adulthood, with the appearance of motor disorders and dementia, for the so-called fragile X syndrome (the second most frequent cause of mental debility after trisomy 21 or Down's syndrome) and cystic fibrosis (the most common autosomal recessive disease in white populations of European origin: 1 in 2000 births). Because of the very large size of the relevant gene, the biotechnological diagnosis of Duchenne's muscular dystrophy falls half-way between the direct and semi-direct methods. This gene was isolated in the autumn of 1985. It can now be tracked down by scan with an error rate of less than 2 percent. But in 30 percent of cases, the anomaly is due to a recent mutation and not to family antecedents.

As can be seen, the diseases traceable through DNA analysis are all extremely serious ones. But they already show quite clearly that the management of one risk - the risk represented by anomalies in development - leads to the formation of another risk - the risk presented to a state or an international community by the installation of a policy for the prevention of these anomalies. As instruments of the containment of this risk, the techniques of pedigree analysis and gene identification, whether or not they use the techniques of biotechnology (and the distinction is sometimes an arbitrary one, as one considers the social dimensions of the question), are transformed in turn into a potential source of risk.

The evaluation of these two types of hazard respectively is rendered particularly difficult by the disparity between their functioning and impact, and the different possibilities of measuring that impact. In one case, the calculation of the risks involved is essentially quantifiable, and made up of an assessment of the cost represented by the maintenance of individuals affected by a disabling condition requiring a greater social commitment, further compounded by the fact that current progress in medicine is extending the lifespan of people who would formerly have been condemned by their handicap and associated problems to an early death (Baird and Sadvonick, 1987).

Similarly, it is easy to observe, as studies in clinical psychology have shown, that the constitution of a sphere of risk concerning abnormalities in development, through the spectacular reduction of the other risk factors affecting the lives of children, gives rise to the emergence, among potential parents, of an ever stronger element of fear and, as one might expect, to a demand for the elimination of that fear. Thus, one may speculate that the development of increasingly early detection tests will lead, among the social group consisting of the totality of potential parents, to a growing unwillingness to tolerate abnormalities which could have been detected early enough for the simple elimination of foetuses carrying these defects.

Many factors come into play in calculating the possible justifications for genetic analysis: the frequency of the disease, its severity, the cost of diagnosis, etc. The other type of hazard is obviously not calculable on the same basis. It is essentially of an ideological nature, and has to do with a society's image of itself and of the position of individuals within it, and also with how this image is transformed within certain categorical imperatives, and the constant adaptation of morality to scientific and technical change.

The tracing of certain diseases poses problems of a very special kind. An autosomal dominant disease, Huntington's chorea, on average affects one out of every two children born to a sufferer. A recent article in Science reported that a doctor in the Medical Centre of the University of Indiana had been contacted the previous year by an adoption agency, for the following reason: a little girl aged two months, whose mother had died of Huntington's chorea, was to be adopted in the near future, but the prospective adoptive parents did not want to take the child before being assured, through genetic diagnosis, that she was unaffected by the disorder. The doctor refused to carry out the diagnosis, on the grounds that only adults can in conscience take the decision to know, or to continue not knowing, that they might possibly have this gene.

This is just one example of the dilemmas created by the recent development of a test for the Huntington's chorea gene. Enquiries have already been carried out among families at risk, to find out whether they wished to undergo this test. The personal, family, career and other problems which will certainly follow from the existence of such a test will probably soon be doubled by the investigations which life assurance companies will want to undertake when contacted by these people.

The problems are obviously not confined to the medical sphere. The liveliest debates sometimes occur when they emerge from that sphere, or threaten to emerge. One very special form of analysis is concerned with what is sometimes called genetic fingerprinting. An article in Nature told the story of a young Ghanaian who wished to join his mother, resident in Britain. (Jeffreys et al., 1985) At first he was refused an entry visa by the UK immigration service, because the usual blood group tests failed to determine whether he was the son or the nephew of this woman. A direct analysis was then made of genetic information contained in his DNA, his claim to be the son of a legal resident of the UK confirmed, and the boy was given an immigration permit. This was one of the first published uses of genetic fingerprinting, a technique which has been patented and is now being commercialised by a company in the UK created specifically for this purpose.

This same technique, used on blood or sperm samples, can make it possible to charge a suspect arrested for murder or rape, by demonstrating the identity of DNA in the suspect's cells with that in cells in blood or sperm found on the victim. In general, this method could be applied to all cases where there are problems of identification, as has already been done in connection with paternity cases. Another form of genetic tracing, which is at present very controversial, has to do with job recruitment. There are already many examples of such applications. Certain American firms have refused to take on healthy carriers (ie non-affected carriers) of one copy of the gene which causes sickle-cell anaemia in individuals who inherit two copies of that gene. The pretext was that this factor made them unsuitable for certain working conditions. Sickle-cell anaemia basically affects blacks, and one may well ask whether, in periods of economic crisis and xenophobia such as we are currently experiencing, this application of genetics to the world of employment is liable to degenerate essentially into a new form of discrimination.

Thus procedures which seem initially just to offer new ways of obtaining genetic information about individuals at risk in some way are, on closer examination, highly problematic, given the likely context of their use.

## NEW INTERVENTIONS: GENE THERAPY AND PREDICTIVE MEDICINE

The question of diagnosis is placed in practice between two other questions: the question of therapy and the question of prediction. At the present moment, we do not know how to cure the diseases which are traced in this way. It is often stated that the first application of genetic manipulation on human beings is imminent. What will be involved is a genetic "graft" into the bone marrow of children affected by serious immune deficiencies (the "bubble children"). The very first attempt at such a genetic transplant was undertaken in 1980 by an American, Martin Cline, on two women affected by thalassemia. Judged premature, this attempt raised a public furore, all the more so as it had simply ignored the opinion of the NIH (National Institutes of Health). Other attempts are about to start today. Opinions are divided at this point, although on balance they are more favourable to the attempt than they were in 1980. Thus one may expect that the judgement not only of the scientific community but also of the various national and international authorities concerned will be largely a function of the results achieved by the graft. If they turn out to succeed, they will be judged to have been acceptable all along.

But this is only one special form of genetic operation, which only affects the person on whom the graft is performed. The debate obviously takes on very different dimensions when one is talking about an operation on the genes of an embryo, which not only transforms the genetic patrimony of the child, but also those of the child's future descendants. The danger, in any event, does not really reside in a possible therapeutic operation. If one supposes that it will be possible to practise this type of diagnosis in vitro, it would be a less logical procedure to inject a healthy gene into a fertilised ovum which had been found to contain an anomaly at the very beginning of the process of differentiation, than to eliminate those ova collected and fertilised in vitro which carry the abnormality, and only implant into the mother's uterus the unaffected egg or eggs. The hazard would obviously consist most of all in a deliberate modification of genetic features hitherto inaccessible to deliberate human intervention, such as height (by injecting growth hormones, as has already been done with mice, turning them into giant specimens) or other features which are luckily protected from temptation by the extreme complexity which makes them up (intelligence and beauty, for example).

The other major subject of controversy concerns what is called predictive medicine. Some writers envisage, for example, the systematic tracing of hypercholesterolemia. An English team has recently revealed the existence of genetic markers for an autosomal dominant disease, polyposis coli, which nearly always develops into cancer of the colon. (Bodmer et al, 1987) This type of test, carried out on children at the age of 18 or 19 - the age at which the signs of the abnormality appear - could lead in certain cases to drastic preventive measures being taken (such as the removal of all or part of the colon).

When the genes causing susceptibility to the main diseases (cancers and cardio-vascular disease) have been isolated, what preventive measures could one envisage being taken? In his article in this volume Daly suggests that this area of human genetics is exciting commercial interest. A prenatal diagnosis of these genes would not make much sense: two thirds of all people die from cardio-vascular diseases at present, while one-third die of cancers, so an effective diagnosis in a preventive context would simply lead to the whole population being aborted. At the moment, therefore, predictive medicine is seen by its proponents as a post natal diagnosis of genes of susceptibility, which must be accompanied by pharmacological, dietary and lifestyle measures, calculated to ward off the development of the diseases in question. These measures are not very different from the ones already known, but are liable to weigh very heavily on a person's life if imposed from the beginning of it.

## ETHICS A LA MODE

One can assume, without any great risk of being wrong, that articles and reports of all kinds on these questions as a whole, including the ones not included on the agenda of the Dublin meeting, will continue to be written about endlessly. Such articles and reports characteristically predict that we have merely arrived on the threshold of this genetic revolution, and that the problems which it will pose for our societies are just beginning. One can also be forgiven, if one stands back and takes an overall look at the mass of this material over the last ten years, for entertaining some doubts as to the intrinsic effectiveness of these studies. For some years there has been a particularly copious output of this type of writing, under the heading of bio-ethics or biomedical ethics. Ethics committees of all types - hospital committees, professional committees, national committees and so forth - have been asked to give their opinion on general questions, or on particular projects for experiments. The progress of some of these meetings, the problems raised there, and the comments which emerge, would sometimes give one cause to think that the important choices do not really derive from the various opinions expressed by these authorities. This is all the more true since, being usually made up of a substantial majority of doctors, who are often personally involved themselves in the experimental projects on which an opinion is being sought, the committees are particularly careful to avoid hindering the steady progress of scientific research. When one considers that many scientific journals are now refusing to publish the findings of research and experiments without these having been previously endorsed by an ethical committee, one may legitimately doubt that the motives which lead doctors and researchers to seek such endorsement are always exclusively of an ethical kind.

I have put forward elsewhere the hypothesis that bio- ethics is the moral equivalent of a nursery for exotic plants: a theoretical space where societies may grow accustomed to the scientific and technical innovations which these societies have produced, and which have to do with human beings. (Fresco, 1986) According to this hypothesis, the explicit content of bio-ethics - questionings of all sorts, qualified endorsement, condemnation without appeal, detailed criticisms, etc. - is not very significant. That would explain, at least partially, the character of the output, which is frequently very repetitive. What counts is not what one says, but the fact that one has something to say on this subject. The various positions taken up are so many mechanisms of familiarisation with a reality which it is important to master verbally - all the more so as one often feels vaguely that the warnings and moratoriums issued do not necessarily have a practical effect on the speed of preparation of these techniques and their further application. Various authorities - professional, union, political, religious, national, international, etc. - feel obliged to contribute to what is called (perhaps for want of a better name) the debate.

There are now professional specialists and teaching programmes in bio-ethics. Recently, there was even a competition in medical ethics, organised by the Journal of Medical Ethics, which offered a prize of a hundred pounds sterling for the best response to the case described (a woman who decides to leave her husband because of his spectacular mood swings after a stroke has left him slightly paralysed on one side). Answers, between 1500 and 2500 words in length, were to be sent to the editors by 15 May 1987. One may be amused, or shocked, by such a competition. One might also see it as just one more sign of the current fashion for ethics, which emerged from the questionings already mentioned.

## FROM QUANTITY TO QUALITY

Besides, one frequently gets an impression of two parallel lines of discussion, apparently having the same objects in view, but not capable of being connected in any obvious way. One of these lines, followed notably by ecologists, certain feminist groupings, and supporters of a radical critique of science, reject a large proportion of the researches and applications currently taking place, as threats to humanity. The other line, followed essentially by doctors and biologists, reiterates its confidence in the quality and respectability of these techniques in the name of the constant and necessary progress of science.

On the one hand, antibiotic treatments and vaccinations have considerably reduced the rates of infant mortality, while on the other hand there has been, since the 1960s, a recognition - de facto at first and later as of right - of scientific methods of contraception and abortion which make it possible now to organise a generally effective control over the quantity of births. This has shifted the attention of the medical profession and the public authorities onto the quality of the children who are about to be born, which in turn has led to the introduction of policies for the prevention of what has now become the leading cause of infant morbidity and mortality: developmental abnormalities. Of course there are individuals and associations irrevocably opposed, for reasons of principle, to the deliberate destrucion of an embryo or a foetus. And one must take account also of the reverse swing that has been observable for some time in connection with the still very recent liberalisation of abortion laws. Still, a social consensus does seem to have emerged that is is desirable to reduce the numbers of births of seriously handicapped children (physically and/or mentally). The considerations stated or implied include concern for the child, concern for society, and concern for the family.

It is quite exceptional, however, that eugenics should be spoken of in this context, as the work is charged with such historical connotations that it only appears in current discussions accompanied by expressions marking the horror which it inspires. Although eugenics was a familiar, accepted concept - even a cherished goal - during the whole first half of the twentieth century, the term practically disappeared from scientific and general literature after the end of the second world war, and the discovery of the scale of the crimes committed by the Nazis in the name of an explicitly eugenic ideology. As early as the 1930s, the discoveries of the newest science of genetics were undermining the scientific pretensions of eugenics, while Hitler's applications of the theory ruined its respectability; the result was that the scientists of the time distanced themselves, for the most part, from theories which had been rejected simultaneously by science and morality.

For some years the term has been surfacing again in debates, almost always in the same guise, branded as a threat, a "spectre", according to the ritual phrase (this is almost always the term used, as for example in the statement by Robert Badinter, who was then the French Justice Minister, at the Council of Europe meeting in Vienna on 20 March 1985 on human rights and medical progress: "A spectre haunts the world, named eugenics, linked to certain totalitarian practices").

The dimensions of that spectre are such that is seems necessary for scientists today, especially geneticists, to put forward various arguments to prove that the application of new analytic techniques, even on a large scale, has nothing to do with questions of eugenics. Their arguments include the following points: the measures proposed may be urged on patients, but are absolutely never compulsory; medical intervention never takes place after the birth (which would be infanticide), nor without the free consent of the spouse or partner. Finally, the methods of diagnosis mostly have an effect which is not eugenic but dysgenic (ie they increase, rather than decrease, the number of abnormal genes in human populations), because before the introduction of these techniques many couples had given up the idea of having more children after the terrible experience represented by the life (and frequently the death) of a badly handicapped child, whereas the new techniques now make it possible to assure the future parents that the child whom they are expecting is unaffected by the disorder in question. However it is likely that such children will be healthy carriers of the relevant gene and may pass it on to the next generation. The obvious relief that one finds among many geneticists, when they can prove scientifically that their activities are dysgenic rather than eugenic in practice, undoubtedly reflects the difficult interaction of moral and scientific points of view. The moral perspective is considerably sharpened by the crimes of the recent past, but also affected by the Hippocratic principle that the doctor's duty is to look after the health of patients here and now, without worrying unduly about the possible effects of his or her consultaions on the genetic future of mankind. The scientific perspective is borne along by its own advances, and justified by its individual clinical applications.

Among the elements which help to make the current state of eugenics into such a complex area, one must give considerable importance to this contradictory double movement of continuity and discontinuity on the one hand between the characteristics of the eugenic era and ideology of the first part of the twentieth century, and on the other, the ideology of prevention as it is constituted today. The second world war and the realisations that went with it, marked the break, and there are very few people today who explicitly espouse a policy of eugenics, as that term used to be understood, although today there are means of implementing such a policy which were not available to the supporters of eugenics in the past. But these means are instruments of a kind of de facto eugenics which, because of recent history, seems to be acceptable only on condition that it is specifically not called eugenics.

The problem now arises for scientific and political authorities in the various countries involved, as to how they should bridge the moral gap between (1) the revulsion that may be inspired by the idea of selecting and eliminating foetuses on the one hand - a collective, abstract idea which, besides the obvious weight of the terminology, bases its revulsion on the excesses, lunacies and mass murders of recent history - and (2) the concern that so long as an effective treatment has not yet been introduced for certain diseases, no individual should be subjected to a life of torture which could have been avoided by preventing him or her from being born.

But what guarantee is there then that the search for such treatments will occupy a sufficiently large place in the health budgets and the choices of indiciduals and societies, in comparison with the simpler and cheaper option of eliminating defective foetuses ? And where are we to draw the line between those considered too defective to be brought into the world, and those who will be authorised to live? And how can one be sure that the maintenance of democratic procedures, where responsibility for such a decision will rest with the parents of the potential child, will bear up under the pressure of such a choice? We may remark in passing that the ultimate barrier which seems to be represented by individual freedom of choice is by no means necessarily a guarantee of maintenance of this democratic procedure - unless one is prepared to adopt a naive and democratic view of the question. Ultrasonographers often refuse at present to ascertain whether the foetus which they are examining has a harelip, because they know that this deformity can now usually be operated on successfully after the birth, and they want to spare the future parents from suffering and worry during the later weeks of pregnancy. But what will happen when this kind of malformation, and others which do not affect life expectancy or the essential functions of mind and body, are traceable at an early enough stage in the pregnancy for a so-called abortion of convenience to be carried out in the timescale envisaged by the law? How will the decision be made by those who have this information, between keeping it to themselves at the risk of being subjected to legal proceedings afterwards on the grounds that they have failed to divulge medical information which belongs by right to the interested parties, and handing over this same decision to the potential parents? Who will be able to claim any real freedom of choice in those circumstances?

The questions raised here are still only (if one may use the phrase) in an embryonic state. If eugenics is involved, in fact, we are still talking only about pre-natal eugenics, intervening between the conception and the eventual birth of a child. For centuries we have been happy to help newly-born children into the world by mechanical means (Caesarian section, forceps). Their condition was observed, and what we choose to call nature decided who was to live and who was to die. And at the risk of being a little anthropomorphic, we know that Mother Nature was often a cruel and bloodthirsty creature, and that in these questions there is no Golden Age to which we can hope to return by means of pious ecological wishes, to escape from the iniquities of progress etc. It is not so long ago that so-called perinatal medicine was introduced, and the moment of birth was no longer seen as a radical break in the continuity of its field of application. The unceasing journey upstream, towards the source of the pregnancy, has subsequently been reflected in increasingly early analytic procedures, right up to the study of the first moments in the embryo's life. But these different procedures for analysis or selection still remain, for the moment, pre-natal but post-conception. Predictive medicine will necessarily involve pre-conception applications, which in suitable cases will make it possible to avoid recourse to an abortion which will seem even more scientifically retrograde than morally problematic.

In 1950 Alfred Sauvy prophetically concluded his preface to Jean Sutter's book, L'Eugenique (Paris, Presses Universitaires de France) with the following observation: "The apparent decline in eugenic science should not fool anybody. In other forms, and perhaps under different names, this science will undoubtedly provide the liveliest and most far-reaching debates in the society of tomorrow." Only the future will tell whether the debates and seminars will concern us today will have an influence, in one way or another, on the course of events, or whether they will merely serve as the moral discharge which a society has to administer to itself, so that it can contemplate the introduction of a system in which technical considerations will be less and less able, through their failures and imperfections, to serve as a barrier against advances which are felt to be threatening, and in which the human race will be less and less able to rely on its own technical limitations to serve as moral boundaries when dealing with changes which are perhaps inevitable.

# PART THREE: FOOD AND AGRICULTURE

# 7 Biotechnology and Food/Agricultural complexes

Pascal Bye

## INTRODUCTION

It is undoubtedly the case that whenever any dramatic change in technology occurs, the real or anticipated impact of the new situation arouses reactions of enthusiasm or rejection. Biotechnology is no exception to this rule. When it was first talked about, at the height of the economic and oil crisis, biotechnology was supposed to help the agriculture of the developed countries to get out of the impasse, by improving their competitiveness on non-food and international markets through a lowering of costs. It was then more common to emphasise the positive effects arising from the introduction of a new technology, than to explore the conditions and consequences of securing these effects.

This optimistic vision of the future gradually got more complicated. Hasty extrapolations of experimental findings had to give way to an analysis of market conditions, and the role played by the major agricultural and industrial forces. Biotechnology appeared as a factor exacerbating competition, both at a national and an international level.

The main requirement today, without sinking into blissful optimism or rabid pessimism, is to examine the possible role of biotechnology as a means of restructuring industrial production and patterns of trade on the one hand, and as a means of redefining the functions of agriculture, on the other. On these questions, one finds a greater number of detailed observations than general analyses. They therefore constitute new fields of research in themselves. This introductory note for the discussions has the following two objectives:

1. To describe the principal effects to be expected from the impact of biotechnology, depending on whether one emphasises the technical importance of the new technology ("Biotechnology to re-launch agricultural growth") or market phenomena ("Biotechnology as a destabilising factor in agricultural activity").

2. To compare these two approaches and highlight the role of biotechnology in redefining the functions of agriculture ("Biotechnology: towards a redefinition of the functions of agriculture").

These three different possibilities will be considered in turn.

## BIOTECHNOLOGY TO RE-LAUNCH AGRICULTURAL GROWTH

The first scenario, based on a set of observations or deductions from evaluations or technical extrapolations, offers a picture of the positive effects which will follow from the introduction of biotechnology. These good effects are all founded on a very noticeable increase in agricultural and food productivity. The causes of this increase are, for example, the improvement or creation of new varieties of seeds; advances in animal genetics and stockbreeding, complemented by the use of growth hormones in animal agriculture; an increase in the protein, carbohydrate or lipid content of plants; improvements in fermentation processes; and the extraction of greater value from by-products. The cumulative effect of these improvements, or "scientific and technical revolutions", would allow the following achievements at Community level for food and agricultural activities. A significant drop in production costs would open up new opportunities. The relative reduction of long-term costs would improve the competitiveness of European agriculture on the industrial and food markets. [1] The portion of usable agricultural land given over to exports would increase, constituting an indispensable aid to agricultural growth as well as food development.

The opening up of markets other than food would be a function of, on the one hand, the fall in the price of agricultural products, and on the other the reduction of the gap between the price of non-renewable and renewable raw materials, and of improvements in techniques for transforming and enhancing the value of biomass. These factors taken together would help to increase the industrial market for agricultural produce. Industrial uses are already believed to absorb 10 per cent agricultural products, and 20 per cent of household expenditure. The limits imposed by the non-elasticity of demand for food in high-income countries would be in this way circumvented..

[1] The proportion of fixed capital expenditure (long and medium term expenditure) would go down in proportion to expenditure for advances on crops (short term).

An improvement in the EEC's autonomy or self-sufficiency as regards food would also result. The application of biotechnology in the production or transformation of food and agricultural products, as well as stimulating exports, would make it possible to diminish imports. Thus, certain recent calculations show that it is technically feasible to envisage the replacement of all animal feed imports (over 20 million tons per year) into the EEC, as well as oil and vegetable fat (over 4 million tons per year), as well as making a significant contribution to cutting imports of forestry products (over 120 million cubic metres per year) and fossil hydrocarbons (about 80 millions of petroleum equivalents per year) used as basic products and mostly imported.

Another result would be the creation or strengthening of economic activities linked to the development of agricultural production and biomass processing, both upstream and downstream. These activities, which are not all concerned with the food markets, will be able to become more varied, less concentrated, and thus contribute to the development of the rural environment.

Although it is possible, on the basis of often experimental findings, to draw up an initial balance-sheet of positive effects expected from the introduction of new technology, this "scientistic" approach sins in at least three ways. Firstly it often extrapolates major "trends" of technological renewal or transformation on the basis of scientific discoveries - breakthroughs in genetical engineering and fermentation are most often cited. Such theories undervalue the slowing-down effects attributable to the social environment, and, more importantly, the voluntary restraints exerted by the forces dominating the major techniques and the markets.

Secondly this approach does not give enough weight to contradictions arising from objectives which compete against each other rather than complementing each other. Thus, it is obvious that agriculture in Europe cannot concentrate simultaneously, even if its international competitiveness improves, on opening up new external markets and on national markets. Similarly it cannot at the same time encourage agro-industrial concentration, so as to benefit from economies of scale, and diversification, so as to ensure the enhancement of the value of all components in the biomass.

Thirdly it minimises the importance of prices and markets, neglecting among other things the role played by supply and demand on prices and the market. It neglects, for example, the impact of the effects of co-production on prices relating to agricultural raw materials, and the more general impact of macro-prices (exchange rates, interest rates, prices of other factors of production), which often involves government policies.

It is precisely these inadequacies that are highlighted in the most recent analyses, which may be described as "market oriented", as they pay more attention to questions of the diffusion of biotechnology, than to questions of its generation.

## BIOTECHNOLOGY: A DESTABILISING FACTOR IN AGRICULTURAL ACTIVITY

The emphasis here is placed on the appearance of three phenomena linked to the spread of biotechnology: new forms of competition, the strengthening of powers upstream and downstream of agriculture, and the definition of a new technological model.

By making it possible to replace one substrate with another, while ensuring a better exploitation of the available biomass, which is a stimulus to industrial and commercial co-production strategies, biotechnology is re-launching international competition in agriculture on the one hand, and competition between different economic activities on the other. The introduction of biotechnology is thus not always synonymous with the opening of new outlets for agriculture. It can lead on the contrary to the reduction or even the destruction of its traditional outlets, thanks to improvements in the techniques for exploiting the biomass. For example, the possibility of introducing new feedstuffs enriched with synthetic proteins considerably reduces the outlets for cereal crops. The recycling of milk whey in animal production has similar implications for vegetable-based proteins. The processing of effluent for food purposes decreases, rather than increases the use of agricultural products for food purposes.

Improvements in the techniques of biomass processing (fermentation; fragmentation-recomposition; new uses for starches; selection of plants rich in protein or able to survive in the available environmental conditions) have multiplied the possibilities of replacing one agricultural product with another, but also of replacing agricultural products with maricultural products (seaweed, for example) or forestry products. The general spread of these techniques also calls into question the restricted specialist category of the agri-food industry which is thus relatively sheltered from competition.

Progress made in the field of fermentation for the treatment of starch has been for example one of the foundations of the industrial production of corn syrups with high fructose content (HFCS). These innovations have encouraged the partial replacement of sugar-bearing plants (cane and beet) by sugars extracted from cereal starch. In addition, the by-products of HFCS production (corn-gluten feed and mill proteins; lipid products) have appeared on markets hitherto dominated by traditional oil-seeds and cereals.

Advanced biotechnology can modify not only production conditions but also market conditions. It will thus be contributing, in already saturated agricultural markets, to a swelling of stocks and a downgrading of price levels.

In addition to these matters concerned with competition and outlets, emphasis is being placed on a second phenomenon: shifts in power within the agriculture/food complex induced by improvements in genetic technology. These changes take concrete shape by simultaneously strengthening the positions of the designers and sellers of new production methods - seeds, feed additives, growth promoters, processing enzymes - and also strengthening the position of those using new products. Furthermore, there is a strengthening of the relationship between agriculture, increasingly compelled to buy its means of production in order to sell products destined for ever more precisely targeted uses, and industry which, by setting up a coherent technology chain, to supply material to and purchase it from the farm, has reinforced its position both upstream and downstream of the agricultural process.

Thus, the introduction of techniques for the production and processing of biological materials would tend to operate at first for the benefit of those countries and companies most capable of making a clear break with the current organisation of agricultural/food complexes. They would replace then the combination of heterogeneous and well-worn technologies used since the industrial revolution, with a set of technologies which are homogeneous but more and more monopolised by the producers of new technological and scientific knowledge. The development of biotechnology in the agricultural/food complex would be one of the principal ways of advancing the power of the new actors on the industrial scene. This would mean that agriculture would give up a little more of its independence. It could no longer hold on to its specific mode of activity, based on the craft-scale application of empirical knowledge and experience not reproducible on an industrial scale. Neither would it be able to choose between different technological models, because of its dependence on the economic or political powers already mentioned.

In order to take advantage of the opportunities offered by this new dominant technology (increased productivity; lower production costs; more effective linkage of activities upstream and downstream of agriculture), agriculture would have to meet a number of new criteria accentuating, rather than reducing, developments widely heralded during the preceding decades. This would involve a further reduction in the useful agricultural area, a decrease in the numbers of species grown or raised, a concentration of production or a reduction in the active agricultural population. Rather than contributing to a revival of all agriculture, biotechnology would trigger a new process of elimination and reduction of existing forms of agriculture. This restructuring would operate through a rigid selection of spaces, species and producers. What must therefore be considered is not the potential opened up by a technology which is "a priori" more productive, but how this potential will be used by existing forces with an interest in introducing it, shaping it or resisting it, in order to reinforce their own positions. The analysis of the impact of biotechnology must include an analysis of the power relationships connected to its diffusion.

## BIOTECHNOLOGY: TOWARDS A REDEFINITION OF THE FUNCTIONS OF AGRICULTURE

A comparison between the two preceding approaches serves to point up three important aspects of the impact of a radical change arising from the adoption of biotechnology. These may be summarised in the following manner, bearing in mind that although there is a good number of studies or observations relating to these issues, they are rarely approached on their own terms, and thus constitute new fields of research.

Firstly the adoption of biotechnology modifies the organisation of production, upstream and downstream of agriculture. Secondly increased use of biotechnology challenges the present use being made of production resources, and especially of land, and helps to change the techniques of agriculture. Thirdly, technological change, the renewal of scientific and economic forces, and the emergence of new openings, all combine to change the traditional functions of agriculture. These are now considered in turn.

The restructuring upstream and downstream of agriculture brought about by biotechnology represents an acceleration of technological change, rather than a real technological revolution. This is the perspective generally put forward by writers considering the expected impact of biotechnology on agriculture. In particular, this impact is predicted to take the following form. Firstly there will be a gradual replacement of the physical/chemical techniques based on post-war mechanical equipment by techniques capable of dealing with greater complexities of production and the exploitation of biological material. This movement confirms the strong position of new industrial forces within agri-supply: seed producers allied to large chemical or parachemical groups; producers of complex chemical molecules used for veterinary purposes or crop protection; animal or vegetable breeders; producers of additives for animal feeds. These firms can modify the whole relationship between industries and activities through biotechnology, which thus becomes not only a complementary resource, but also a replacement for other processes.

Secondly one can predict a challenge to the organisation of agriculture and food processing in a specialist sector. Upstream, the adoption of biotechnology leads to stronger links between engineering, chemistry and biology. The aim is less to produce in ever-increasing quantities than to produce at better cost, and with a view to downstream openings, which are no longer purely concerned with food, but also with energy and chemical applications. Downstream, we are finding a progressive challenge to established production processes. One no longer sees an increasingly specialised product derived from agriculture, undergoing successive transformations in order to become a specific food product. What must be envisaged, on the contrary, is that a complex product arising from agriculture or other biological or physical settings could be decomposed and then reconstituted to make a large new range of non- specific products.

Thirdly one can envisage a multiplication of the relationships between acitivities upstream and downstream of food-processing. These relationships, considerably extended by the fact that the forces acting in the agri-food complex are heterogeneous ones, would give rise to higher production costs and progressive saturation of outlets. Increased use of biotechnology both upstream and downstream of agriculture would mean that the industrial forces involved in agri- supply of transformation would wish to ensure at the same time a coherent use of technology and a control over new processes of production. This may already be seen clearly when a fruit and vegetable canning company invests in research into new seed varieties which could decrease its processing costs, or a chemical firm specialising in the production of food additives enlarges its activities to include animal breeding, so as to achieve greater control over its outlets. Developments of this sort are not without consequences for the technological approaches followed within agriculture.

The restructuring of industry upstream and downstream of agriculture has direct consequences for the orientation of agricultural technology. Among the positive effects are the rationalisation of investments, greater flexibility and the use of intermediate products, which can bring rapid variation in the volume of production, lower unit costs, linked to increases in productivity and the exploitation of by-products. The fact remains that opinions vary when one is discussing the effects of biotechnology alone.

While the introduction of biotechnology undoubtedly improves the performance of agriculture overall, it reduces the ability of farmers to breed for variety in plants and animals. Biotechnology increases their dependence on the industries producing and selling these means of production, all the more since their use induces the purchase of other industrial products. Although they are the only consumers, farmers can determine neither the parameters, prices nor conditions of use of these new "technology packages".

The introduction of biotechnology theoretically makes it possible to exploit new agricultural land, essentially because of the new development of new plant varieties, which survive in different climatic conditions and can tolerate agreater degree of enviromental stress. Thus the use of new seed varieties induces more and more purchase of precise ranges of plant health products; the use of particular new breeds of animal induces the purchase of certain food or veterinary products. Biotechnology makes it possible to exploit vegetable or animal products which have not been marketed, or have scarcely been marketed, up to the present. In theory it makes possible the use of biomass products, which are excluded or not used at the moment, as genetic engineering or fermentation processes make it possible to use a small range of substrates to produce a large variety of finished products. In fact, rather than extending the boundaries of agriculture, the adoption of biotechnology reduces it to those areas which can produce the most versatile species of animals and vegetables at the lowest costs. Biotechnology would tend to encourage intensive rather than extensive farming, and a concentration of products and producers rather than diversification.

The forces involved in spreading biotechnology would tend to encourage this trend, so as to increase their sales at the same time as controlling the prices of their own raw materials. The diversification in the range of products derived from the biomass would benefit those who process the biomass rather than those who produce it.

The use of biotechnology goes in the same direction as a movement already begun in agriculture, which consists in the gradual replacement of long term investments in land, and in equipment linked to extensive agricultural production, by medium and short term investments linked to intensive and "overground" types of production. In concrete terms, investments in land and specialised machinery to exploit large tracts of land would be replaced by investments in irrigation and buildings, and in supervision and control equipment, linked to the introduction of new procedures.

## BIOTECHNOLOGY AND THE WIDENING OF AGRI-FOOD FUNCTIONS

Designed and perfected in order to improve agri-food production, the physical and chemical technologies applied in the agricultural systems of the industrialised countries have broken agricultural production processes down into many different operations, ranging from the supply of production resources, to transport, to processing, to packaging and distribution. As we know, this widening effect has already encouraged the extension of industrial outlets, especially for the mechanical engineering and chemical industries, helped the proliferation of specialist industries downstream of agricultural production, stimulated the creation of service activities and increased exports. Various characteristics of biotechnology, in particular the ability to use a succession of different substrates and to broaden the range of intermediate and finished products, modifies the functions traditionally entrusted to agriculture in developed countries, and especially three of these functions.

The emergence of new openings upstream and downstream of agriculture goes in the same direction as developments found in certain industrial sectors. Agriculture is becoming one of the foundations of growth in these sectors, through biotechnology. It is considerably extending the new opportunities, as new areas of experimentation open up in the production and utilisation of biological material. It extends the range of intermediate and finished products. But it also contributes, by the same token, to the restructuring around these chemical technologies of the whole range of agri-food activities. This new pivotal role attaches to industrial activity based on organic chemistry, because of the ability of firms to act simultaneously as suppliers of goods for production and as users of products emerging from industrial activity.

The re-grouping of agri-food activities around chemical and parachemical production is going to bring noticeable changes in the traditional functions of agriculture. Non- food outlets should increase, simply through the closeness of industrial relationships between the large groups involved in organic chemicals, petrochemicals and pharmaceuticals, but also their recent economic and commercial reorientation. This development is however closely dependent on the decrease in the price differential between products coming from agricultural biomass and the price of other feedstocks. It may be expected that the introduction of biotechnology will increase that differential.

The food function in the strict sense would be increasingly overtaken by a function linked to nutrition and health. Food budgets and household budgets would become much more closley linked than previously. Agri-food production could benefit from this stimulating effect, if the range, and more importantly, the composition of agricultural products were noticeably modified. Biotechnology could participate in this development.

The production of basic raw materials of vegetable or animal origin would be carried out in ever smaller spaces. Biotechnology should however contribute to the enhancement and maintenance, for purposes other than agriculture, such as leisure, environmental conservation and forestry, of land freed from agricultural production. The new functions, hitherto associated very closely with agricultural production, will involve the implementation of new techniques, and here biotechnology has a role to play.

The implementation of these new functions effectively challenges the role of agricultural production. By becoming less specialised, its organisation will change, as will the features of the production resources it uses and its location in physical space and in the economy. These same developments bring agriculture closer to a type of chemical industry, which has been working for a long time, not in terms of product but in terms of function, not in terms of categories but in terms of industrial systems, not in terms of boundaries but in terms of adaptable spaces, in relation to international strategies and the redeployment of resources.

# 8 The Impact of Biotechnology on European Agriculture

Gerd Junne, Jos Bijman

## INTRODUCTION

This paper will survey the impact of biotechnology on European agriculture. Given the restrictions with regard to space and the intention to be nevertheless as concrete as possible, the paper will concentrate on three important subsectors: dairy, sugar and starch production. It will describe some of the social consequences of applications of biotechnology (on employment, regional inequality, trade conflicts, relative power of trade associations) and try to indicate the relative weight of the different developments as well as possible trends. The paper will finally present some options for political response at the level of the European Community.

Applications of biotechnology to agriculture will be more important to the European Community than applications in any other field, partly because the consequences of applications in agricultue are very far-reaching, partly because agriculture is of such tremendous importance for the budget of the European Community and European policy making.

Biotechnology will more rapidly find applications in livestock farming than in crop production. This will affect meat and dairy production, of which the latter is of much greater economic importance in Europe. The dairy sector is the single most important subsector of agriculture, accounting for almost one fifth (18.9 per cent) of the total value of agricultural production in the EEC in 1984. This paper will therefore deal with this sector first.

Next the impact of biotechnology on sugar production in Europe will be discussed. Beet sugar in Europe has not been substitued by isoglucose on a similar scale as imported cane sugar has been in the United States. The beet farmers' lobby has been strong enough to keep isoglucose production in Europe within very narrow limits. This example demonstrates how the implications of new technologies depend not only on the availability of new technologies and the economics of their use, but also very much on the political influence of different social groups, a point also emphasised by Byé in his contribution to this volume.

The political power of different interest groups, however, may itself change as a result of the introduction of new technologies. Since different plants increasingly can substitute each other as raw materials for all kinds of processes, different groups of farmers feel more competition from each other, as their crops become alternative sources of the same substances. This competition may lead to new cleavage and contradictions within farmers' associations which can weaken their lobbying power in the future. A decline in the power exerted by agricultural interests in Europe may facilitate a thorough revision of the Common Agricultural Policy (CAP) of the EEC and eventually lead to radically altered patterns of international trade in agricultural products. This in turn may produce new pressures on markets and further stimuli to innovate in particular directions.

## IMPLICATIONS OF BIOTECHNOLOGY FOR THE DAIRY SECTOR IN EUROPE

The dairy sector may be divided into primary production and industrial processing. While primary milk production is the most important agricultural activity in Europe, milk processing is the largest branch of the food industry. Biotechnology traditionally plays a leading role in the dairy industry. Three quarters of the raw materials used by the Dutch dairy industry are converted into various end products by means of microorganisms. A characteristic of the new biotechnology is the ability to manipulate these microorganisms in a way that processing can be better controlled, be done more efficiently and more cheaply, and will give rise to new products.

In the primary production sector, on dairy farms, biotechnology is mainly applied to improve breeding. While there is still some delay in implementing genetic engineering in plant agriculture, great changes will shortly take place in animal production as the result of introducing biotechnology. These changes are taking place in at least four areas: (1) progressive upgrading of breeding; (2) improved food conversion; (3) use of growth and other hormones, and (4) progress in veterinary science.

While progress in upgrading animal breeding has already led to continued rises in productivity, biotechnology can speed up the developent process considerably. In addition to transplantation, in vitro fertilisation and the production of chimaeric animals make more specific selection of genetic characteristics possible. (SRI, 1984) Milk production can also be increased by the use of isoacids to promote the growth of bacteria already present in the rumen, so that the food is better digested and utilised. Improved food conversion is the result. The use of growth hormones such as somatotropin can further increase the milk production per cow by as much as 30 per cent. (Kalter, 1985) Animal care can be improved as a result of applications of biotechnology in veterinary science. Diagnostic kits for animal diseases, using monoclonal antibodies, and new vaccines are already on the market.

The combined use of all these new techniques will lead to a tremendous increase in productivity. In the Netherlands for example where the present average annual milk yield per cow is over 5,000 kg, it is expected that yields will increase to 8,000-8,500 kg by the year 2000. (Scheer, 1985) If the total production volume is not allowed to increase, given the fact that the EEC has already to store more than 1.4 million tons of butter and more than 1 million tons of milk powder and that the butter regime alone cost DM 5.23 billion in 1985, the increase of productivity would imply that one third of the stock would have to be taken out of production. Even if growth hormones were banned within the EEC, a reduction in the number of cattle number by 22 per cent will have to take place. This would mean that by the year 2000, many cattle farmers would have to have left this sector. If developments are left to market forces, a strong tendency towards scaling up and specialisation will result, forcing many small farmers to give up farming.

How will these changes affect different regions of Europe? The productivity increases are unequally distributed throughout the Community. The annual increase in the yield per cow in the years 1973-83 was highest is Denmark and Ireland (2.4 per cent and 2.3 per cent respectively), and lowest in Belgium (0.7 per cent only), where the yield per cow is considerably below the European average. Major differences in production structures are evident from the numbers of dairy cattle per farm. The UK is clearly the leader with an average of 57.1 cows per unit. The Netherlands follow with 40.2. The smallest units are in Greece (3.1), Italy (7.2) und Germany (13.6). (CEC, 1986) Since large farms will be more able to introduce new technologies one might expect a shift of milk production to countries with large units and a large yield per cow. (UK, Netherlands, Denmark), to the detriment of countries with small units and a lower yield per cow (Italy, Belgium, Germany).

Such a conclusion however does not take politics into account. The rise in production in the past ten years has meant large surpluses of dairy products within the EEC. The surpluses cost the EEC large sums in intervention payments, storage costs and subsidies on sales. Exports on the world market offer little relief, since the EEC already accounts for 60 per cent of world trade in dairy products. In 1984 the European Commission therefore decided to introduce production restrictions. A quota system was introduced, under which each Member State was assigned a maximum quota of milk to be produced. Such a system is of fundamental importance for the distribution of the effects of increased productivity. Under a quota system, productivity increases will force producers out of the market where the increases are the *highest*, whereas unbridled market forces would lead to a decline of producers in regions with the *lowest* productivity increases. The quota system thus limits the *inter-regional* concentration process. Concentration will take place instead *inside* individual regions.

The protection of farmers in the less productive regions, however, is relative only. The tendency towards concentration comes in by the back door: via the dairy industry. Developments in biotechnology are facilitating the continued automation of the dairy industry and ensure a longer shelf-life for its products. They therefore promote large-scale, highly automated production of dairy products for delivery to regions which at present are still being served by smaller industrial units. Large dairy plants will of course be set up in regions where milk production is high. And milk production is in fact only profitable if it can take place within a reasonable distance from the factory, because transport costs otherwise become too high. Concentration in the dairy industry, therefore, will also stimulate concentration in agriculture itself, especially in areas where the structure of production leads to high transport costs.

This concentration process would speed up if milk production came under additional pressure as the result of an increasing substitution of vegetable proteins for dairy proteins. Intensive competition in the food industry has meant that raw materials are in demand which can replace expensive products of animal origin in the same way as a large proportion of animal fat (butter) was replaced by vegetable fat (margarine) at the beginning of the century. Now the dairy industry is fearful because of the advent of vegetable proteins. (Van Kasteren, 1985) Where dairy proteins have traditionally been used, vegetable proteins are increasingly taking their place. Chief among these are soya proteins. Well-known examples are coffee creamers, instead of condensed milk, and imitation cheese. The tendency towards more processed food simplifies the exchange of ingredients. Soya and other vegetable proteins are still at a disadvantage when it comes to taste and colour. But biotechnology offers the possibility of doing something about this. The use of biotechnologically developed flavours and processed foods can overcome the poor taste and texture.

A study carried out at the Technical Change Centre in London investigated the potential effects of producing casein (dairy protein) by vegetable means. The substitution of vegetable proteins for dairy protein has remained limited until now, because the specific properties of casein are required in many foods. However, if casein could be made by plants, the substitution would be feasible. As a result of progress in genetic manipulation, the genes which are responsible for casein production could be introduced into plants. Casein from plants would be cheaper than that from milk.

Although this is still only theoretically possible, one can nevertheless calculate the potential effects on dairy farming. If vegetable casein, manufactured from rapeseed, was used for all the cheese produced in the EEC, this would have the following effects on agriculture: 3.2 million cattle, or 13 per cent of the total herd, would no longer be needed, over and above the 22 per cent already mentioned above as probably becoming redundant; 2 to 2.5 million hectares of land now required to feed the above 3.2 million cattle would no longer be needed; and about 2 million tons less of feed grain would be sold less annually. (Technical Change Centre, 1984)

In the present situation of huge surpluses of milk and milk products within the Community, the dairy industry will need to seek alternative applications for milk components and by-products. Biotechnology plays an important role here too. Most attention is being given to the use of whey, a by-product of cheese production. Whey contains a number of useful substances, lactose and proteins being the most important. Whey proteins have good functional and nutritional properties which makes them suitable for use in the manufacture of all kinds of food.

Lactose can be sold in small quantities to the pharmaceutical and chemical industries. It can be converted in an enzymatic process either into a mixture of glucose and galactose, which are useful sweetneners, or to lactitol. This sweetener has only 35 per cent of the sweetness of the same weight of sucrose. Since lactitol is only partly digested by human beings, it has only half the calories of sucrose. The most important use of lactitol is therefore as a bulk ingredient in food in which it replaces sucrose because of its low calorific value. In all cases, however, sweeteners made from milk would have to compete with other sweeteners. Only in a few cases, will their use be economic. Milk components as a raw material for non- traditional uses will not therefore resolve the problem of milk surpluses.

## SUBSTITUTION OF BEET SUGAR BY ALTERNATIVE SWEETENERS

Sugar producers in the EEC are confronted with problems similar to those experienced by the producers of milk and dairy products. Biotechnology makes a double contribution towards increasing surpluses. On the one hand it contributes to higher production levels and on the other hand to the development of new sweeteners which are narrowing down the markets for sugar made from sugarbeet (sucrose). As a result, many sugar-beet farmers will be forced out of production.

The contribution of biotechnology to increased productivity in sugar-beet growing will not be described here, because it is not specific to the EEC. With regard to the development of alternative sweeteners, however, a special situation does exist in the EEC.

While the substitution of isoglucose for sugar in the United States has made tremendous strides, the production of isoglucose is restricted by quota in the EEC. When the starch industry succeeded in making the production of isoglucose profitable, it became a direct competitor of the sugar industry. At that time isoglucose manufactured from starch benefited from both the production restitution on starch and the price guarantee on sugar. This double benefit, however, did not last long. Due to a powerful lobby, a quota system for isoglucose was introduced in 1979. Since 1981 this quota system is part of the EEC sugar policy. As in the case of sugar, there are A and B quotas for isoglucose. Both enjoy price protection, but a production levy must be paid on the B quota. C isoglucose must be sold on the world market without financial support. For the period 1981-1985, the production maximum for A and B quotas was 185,085 tonnes annually.

Crott has calculated the effect which the substitution of isoglucose for sugar would have if there were no production restrictions. He arrives at a replacement of 800,000 tons of sugar annually. (Crott, 1981) At an average sugar yield of 5.51 tons per hectare, this would mean the substitution of 145,191 hectares of sugar-beet. If we take the current EEC (Ten) average yield per hectare of 7 tons of sugar as our basis, the production of 800,000 tons of isoglucose would replace 114,285 hectares of sugar-beet. This is 6.6 per cent of the total area under beet cultivation in the EEC in 1984. The producers in less favourable production regions would be the chief victims. The decline in the number of beet growers would thus be accelerated.

Crott's estimate of the quantity of isoglucose which would be produced if no quota existed is probably too low. In North America, 30-45 per cent of the sugar market was taken over by isoglucose in the space of ten years. If we take the EEC percentage as 30 per cent, we arrive at a substitution of 2.8 million tons of sugar. This would represent over 400,000 hectares of sugar-beet, or the entire sugar-beet acreage of West Germany.

It is doubtful whether the production quotas for isoglucose in the EEC will remain at their current low level. It is true that the lobby from Euopean isoglucose producers is not particularly vigorous. These were for a long time mostly subsidiaries of American companies, with an important role accruing to the Italian concern Ferruzzi only recently. Yet the enormous grain surpluses are demanding new markets. One of the future outlets could be the production of isoglucose based on wheaten starch. Wheat and sugar-beet producers would thereby become direct competitors.

While the production of isoglucose is still restricted, there are other sweeteners produced with the help of biotechnology. Their production cannot be limited in the same way. These sweeteners are synthetic sweeteners such as aspartame and acesulpham-K, and intensive sweeteners such as thaumatin derived from African berries like the "miracle fruit". Intensive sweeteners contain substances that are more than 1000 times sweeter than sucrose. European companies such as Unilever and Tate and Lyle are engaged in researching these substances and are attempting to produce them industrially. (Ruivenkamp, 1986) Aspartame is marketed by the American firm G.D. Searle, recently taken over by Monsanto. Aspartame is expected to command the largest portion of the market for artificial sweeteners. Acesulpham- K is also 200 times sweeter than sucrose. The chief producer is Hoechst in West Germany. This company produces approximately 1000 tons annually. Acesulpham-K is also expected to become an important alternative sweetener. The new artificial sweeteners and the intensive sweeteners will probably take over an important part of the sweetener market from sugar-beet in Europe.

Sugar beet has traditionally been an important crop in the EEC. It accounts for 5.4 per cent of the total value of vegetable products. (CEC, 1985) This figure does not however give an accurate impression of the importance of beet-growing. For many farmers sugar-beet is indispensable as a rotation crop grown as a necessary complement to grain.

The chief production areas are Belgium and the Netherlands, Lower Saxony in Germany, the west of the island of Sjaelland in Denmark, Picardie and the area surrounding Paris in France, East Anglia in the United Kingdom, and Marche, Emilia-Romagna and the Po Basin in Italy. (Lee, 1985)

France and Germany are the largest producers in the EEC. Between these two, there are considerable differences. In Germany, and in Italy, a large number of farms are engaged in beet growing; however, a relatively small average area per farm is under sugar-beet. The most favourable growing conditions are found in France, where large-scale cultivation has economic benefits. If sugar production were left to the free market, production would be concentrated even more in France. If there is increased competition from other sweeteners, it is probable that Germany and Italy will be the first to lose beet acreage. A rough estimate indicates that beet-growing in the EEC gives employment to 375,000 people, 144,000 of whom are in Germany. If there were a drastic reduction in beet production in the least productive units, a large number of these jobs would be at risk.

However, how the effects of a declining sugar-beet acreage would be distributed does not depend primarily on the productivity of the farms concerned but on the political distribution of the production quota per country. The market regulation for sugar is considerably different from that for milk in that it is almost completely self-financing. "A" and "B" sugar are subject to price support ("B" sugar at a much lower level), but a production levy applies to both the A and B quotas. The larger the surpluses the higher the levies. This amounts in fact to an indirect price reduction. The quota system merely ensures that the production of small farmers is not taken over by larger farmers in <u>other</u> countries. While international redistribution of production is slowed down, a restructuring at the national level goes on.

The rapid expansion in the production of alternative sweeteners and the rise in productivity in beet-growing itself will increase the oversupply of the sweetener market, which must result in the raising of the levies, a decrease of quotas, or a change-over to much greater financial support for beet-growing than in the past. In view of the

Community's financial situation, there is little prospect of the last alternative. If merely a further raising of the levies is decided, the restructuring will be confined to a redistribution on a national scale. Should the production quotas also come under discussion, this will open the way to an international redistribution of sugar production within Europe.

## CHANGES IN THE SOURCE OF CARBOHYDRATES

The production of sweeteners based on starch shows that various sources of carbohydrates are increasingly competing with one another. With the growing interest in biotechnological production processes, the demand is growing for starch which can be used as a basis for a large number of products. However, the use of biotechnology has not only caused the demand for starch to rise, but also contributed to the increasing interchangeability of the various types of starch. (Lewis and Kristiansen, 1985) This will place the second largest source of starch in the EEC at risk.

The chief raw materials for the starch producing industry are maize, potatoes and wheat. Maize is traditionally the most important raw material in starch production in Europe. That it is not wheat, which is available in abundance in the EEC, but maize which for the most part has to be imported, which is the most important raw material arises mainly because of the low price of maize on the world market. This low price led before the establishment of the Common Market to the setting up of large-scale production capacities for maize starch. European maize is however becoming more and more important to the starch industry which already processes over a million tons of French maize. (Rexen and Munck, 1984) Generally speaking, the climate of the more northerly EEC countries is not ideal for the production of maize. New varieties have however been developed, which are better suited to this climate and the 'maize frontier' is gradually shifting northwards. The application of biotechnology may speed up this shift.

Next to maize, the most important raw material for the production of starch in the EEC is potatoes. In the Netherlands the production of potato starch is particularly important. The Dutch starch industry is the largest in Europe and the potato starch industry is in fact the largest in the world. The industry is organized by a cooperative (AVEBE) which accounts for 100 per cent of the potato sharch produced in the Netherlands, 65 per cent of the EEC production and approximately 30 per cent of world production. Based on an average harvest of 2 to 2.5 million tons of factory potatoes, this Cooperative produces from 450,000 to 500,000 tons of potato starch a year. (Schogt and Beek, 1985, p. 70)

The use of starch and starch derivatives may be roughly divided into two areas of application: the food industry (about 55 per cent) and technical applications (45 per cent). The technical uses are the most important in the case of potato starch; approximately 70 per cent of the total is used for this purpose. The remaining 30 per cent is absorbed by the food industry. (Ibid, p.79)

Potato starch has advantages over other starch types for technical applications. The higher molecular derivatives of potato starch are quite different to those of grain starches. For the production of low molecular derivatives (glucose and oligomers) the type of starch is not very important. Potato starch is at a disadvantage on this market. The biotechnology industry, which makes the use of more starch possible, mainly processes low molecular starch derivative. Any expansion in the industry will therefore be mainly based on maize and wheaten starches.

The use of wheaten starch in the EEC has expanded rapidly during the last years. Whereas in 1978/79 only some 169,000 tons of wheat-based starch were produced, by 1983/84 this had risen to approximately 360,000 tons. One important reason for this rise is the production of wheat glutens. In the past, European wheat often had to be supplemented by imported wheat for processing in the milling industry in order to increase the protein content. This is why wheat is imported in spite of the EEC wheat surpluses. To avoid the import of wheat, European wheat with a low gluten content is split into gluten and starch. The gluten is supplied to the millers to increase the protein content of flour and now replaces imported wheat with a high gluten content. The wheaten starch is then left over for other uses. A second reason for the rise in the production of wheaten starch is the relatively high production restitutions for this product. (Ibid, p. 71) There is also the possibility that the European maize starch industry may obtain a higher quota for the production of glucose by changing over to local raw materials.

The production level of potato starch can only be maintained thanks to a balance premium paid by the EEC for potato starch. This premium is intended to compensate for the higher production costs of potato starch compared with those of maize and wheaten starches. This support is essential to the producers, since the potato starch industry has a number of structural disadvantages vis-a-vis the grain starch industry. In the first place the raw materials are more expensive. In contrast to nearly all other agricultural crops in the EEC, the productivity of factory potatoes has barely increased over the last ten years. An intensive plan of cultivation involving rotation every other year ran into problems of potato diseases and eelworms. These involve expensive disinfecting procedures. The cost of the raw materials for potato starch will therefore not fall, while grain is becoming cheaper all the time. In addition, the production of potato starch entails the extra high cost of investment in waste water treatment plants. The third disadvantage is the seasonal nature of production which leaves the factories idle for most of the year.

Only financial support from the EEC can compensate for these disadvantages. This makes the industry very dependent on political decisions in Brussels. It was only by intensive political lobbying that the Netherlands, supported by Germany and Denmark, were able to keep the premium at its current level. This support however may weaken. Potato starch might in the future be replaced by barley starch. For some purposes for which up to now potato starch has been preferred, barley starch could just as well be used. Barley is new as a source of starch and its development is still in the experimental stage. The great genetic fexibility of this crop means that it is quite possible to improve the characteristics of starch. This would make barley a formidable competitor to potato starch and would undermine the political support for the balance premium. Barley is a main crop in Denmark, where it occupies 51 per cent of the arable land. (Lee, 1985, p. 77) If barley becomes an alternative source of starch for purposes for which until now potato starch has been used, the Danish famers may no longer be in favour of upholding the balance premium on potato starch.

## THE FUTURE OF THE COMMON AGRICULTURAL POLICY

An analysis of starch production emphasises yet again what became evident already from the analysis of the potential implications of biotechnology for the European dairy and sugar industry. Forecasts of the impact of the use of biotechnology in European agriculture must only be made with great caution. Such forecasts ought not to be based exclusively on an analysis of technological development. All that is technologically achievable naturally does not always come about. Changes do not depend only on what can be achieved in the field of technology, nor on what is economically viable, but to a large extent on which policies can be carried through. An analysis of the political staying-power of individual groups of producers will therefore have to form an important part of any analysis of the probable impact of the use of biotechnology. This impact is dependent on the protection which the European market for agricultural products will still enjoy in the future, but also, for example, on the relative power positions of the sugar-beet, grain and potato producers respectively. Their positions, and those of their 'confederates' in the processing industry and among the consumers, may often be the deciding factor.

It seems certain that it will become more difficult in Europe in the future to formulate a common policy for all farmers. The divide between the grain, sugar, potato and dairy producers will only become wider as their products compete more and more directly with each other as a result of their interchangeability as raw materials, brought about by the development of biotechnology. As a result of increasing internal divisions, the representation of agricultural interests as a whole may be weakened. This would reduce the chance of maintaining the Common Agricultual Policy in its current or comparable form. However, as the protection of the European farmers against worldwide competition is broken down, they will be more and more exposed to the impact of the introduction of biotechnology elsewhere, which will cause the number of jobs in European agriculture threatened by this development to increase considerably.

## OPTIONS FOR POLITICAL RESPONSES

The considerable increase of productivity in agriculture, to which biotechnology undoubtly contributes, will exacerbate the already existing problems of agricultural policy and enforce some radical changes, if these are not already brought about before the impact of developments in biotechnology really start to be felt.

The basic options are either to reduce production or to devote land to the production of alternative products for which surpluses do not yet exist. Even if present production levels are kept up or increase slightly, employment in agriculture will decline as a result of increasing productivity and the concentration of production in the most productive regions. This will in any case cause challenges in the field of <u>social policy</u> and <u>regional policy</u>. A thorough analysis of the expected shifts in production patterns as a result of the application of biotechnology should help to identify the most affected regions as early as possible in order to stimulate alternative economic activities in these regions.

The option to cut down or limit agricultural production could be carried out in two ways. Production levels could be frozen, or reduced, without any changes in the methods of production. Alternatively, such a reduction could be brought about by the gradual introduction of alternative forms of agriculture which are better compatible with a healthier environment. Biotechnology could make a contribution to such a development, for example via the development of crops that would need less fertilizers and pesticides, since nitrates in water and pesticide residues in food are causing increasing concern. The Community could support such a development of stimulating research in this direction and by promoting ecological agriculture via its own activities and publications.

Any reduction of the area under cultivation immediately leads to the question to what kind of alternative use the soil should be put. Part of it might become forest again, and other parts may be devoted to recreational activities, needing some kind of landscaping. A lot of part-time employment could be involved in this kind of activity. In general, the European Community should take the special problems of part-time farmers more into account. Their income might be increased by integrating into their agricultural production activities which add significant value to conventional agricultural products, eg. by having transgenic animals produce drugs with their milk.

As an alternative to the limitation of production, the Community could shift to the production of either more agricultural products than are actually imported, or increase agricultural production for non-food uses, such as raw material for industry or biomass for energy production. The Community will certainly proceed some way in the direction of direct import substitution. But this will invariably increase international trade conflicts, especially with the United States. An increase of the production of energy and raw material for industry would also contribute to import substitution, but the sources of imports would be much more dispersed, and few countries exporting to the Community would be massively affected. For these countries, a fair compensation should be negotiated, comparable to the scheme negotiated with Thailand to phase out the export of cassava.

According to Umberto Colombo, the potential of cultivable land in Europe which could be devoted to energy biomass, is over 20 million hectares. Biomass production could be in the order of 100 million tons of coal equivalent a year, equal to 30 per cent of current European coal consumption, or even more if marginal land were to be brought into cultivation. (Colombo, 1986, p. 11) Rexen and Munck have presented an elaborate plan for the creation of agricultural refineries which would separate the different components of the harvest, some of which would be used for food processing and others as agricultural raw materials. (Rexen and Munck, 1984) The share of natural oils and fats as raw material for the chemical industry, eg. raw materials by industry, could be increased considerably. To realise a higher consumption of agricultural raw materials by industry however, agricultural and industrial policy would have to be coordinated much better than in the past. (Sargeant, 1987)

# PART FOUR: WORK AND EMPLOYMENT

# 9 Qualifications and Training for Biotechnology

Guiseppe Lanzavecchia, Danielle Mazzonis, Anna Luise

## BIOTECHNOLOGY: CURRENT SITUATION AND FORECAST TRENDS

In 1973, the first experiment to introduce extraneous pieces of DNA into a micro-organism from another species generated immense scientific excitement. Industrial and financial interest in the commercial applications of the techniques involved soon followed. It had at last become clear that Watson's and Crick's discovery of the DNA double- helix in 1953, and the subsequent deciphering of the genetic code in the early 1960s, had given birth to a "new biology" which would have enormous consequences not only for the health sector but also for agriculture, food production, energy and the environment. Finally molecular biology was providing not just powerful new concepts, with which to think about biological processes, but also new industrial technologies.

The wave of innovations that followed cannot be attributed solely to our understanding of genetic mechanisms and how to manipulate them. Another contributing factor has been the fact that during the same decades much more has become known about processes in a wide range of cells having many different functions in a wide range of species. This has led to advances in fields such as immunology, oncology, plant physiology, vaccine production, neurobiology and pharmacology.

In the last few years, we have seen a boom in scientific companies originating in the academic world and going on the exploit the commercial potential of technologies discovered in basic research. Their success encouraged the belief that the speed of the phenomenon might be unprecedented. It was sometimes thought that the transition from pure research to applied research, in the development and commercialization of new products and processes, would in this case be almost immediate, and moreover would involve relatively modest investment. This theory has not been borne out in practice, and biotechnology is still, at least in many sectors, more of a promise than an established commercial fact. Industrial experience has shown that, as in other "high tech" areas, the biotechnological breakthrough depends on a combination of many scientific, technological, administrative and economic factors which can be considered only in terms of a future measured in decades. It is difficult to consider these factors separately, product by product and process by process, without having to draw on the critical mass of diverse but convergent experience characteristic of all major projects.

Consequently, rather than giving rise to an improvised and instantaneous revolution, as it were, biotechnology is penetrating only by diffusion into established production sectors, where it is used to increase productivity, to economize on both raw materials and energy, even if only because it permits work at ambient temperatures and pressures, and to improve the quality of the products. The techniques concerned include chemical transformations with biological catalysts, fermentation processes which, although extremely conventional in themselves, use selected or genetically manipulated micro-organisms, the production of new seed varieties or plant clones by selection in tissue culture, the use of single cell proteins as food, the large scale production of alcohol as fuel, and biomass fermentation. Biotechnology does not form part of any radically new industrial programme. It is still something less than a genuine new industry in itself, parallel to the other sectors, with a specific market of its own and a self- generating growth capacity.

Biotechnology should not therefore be thought of as an industry in the traditional sense of the word, but as the use of a biological system to obtain a specific product, or the use of biological techniques, such as fermentation, cell and tissue culture, to create a product, process or service. It is used in a large number of scientific disciplines, including immunology, enzymology, bio-electrochemistry, industrial microbiology and molecular genetics; it has a wide field of technical application, not simply genetic engineering, with which it is frequently equated.

Various industries now have recourse to biotechnology: they include the food industry, to produce additives or nutritive substances, such as cheese, beer, vitamins etc.; the pharmaceutical industry, to extract substances such as insulin or plant-based analgesics from animals or plants, and the chemicals industry, to obtain organic acids, amino- acids and pesticides. Agriculture also relies on biological techniques such as the breeding of new crops or the artificial insemination of livestock, the transfer of embryos, the sterilization of male insects and the use of bacterial or viral insecticides to eliminate pests. Seed selection and livestock breeding are activities dating back thousands of years, and techniques such as the inoculation of micro-organisms into the soil or to make silage were already being used in the early years of this century.

In addition, the instruments used in exploiting the new techniques of biotechnology are based on both old and new techniques of chemistry, physics, biochemistry, molecular genetics, botany and microbiology. For example, separation techniques are based on the principles of chromatography established at the start of this century; genetic engineering still uses very traditional techniques, such as fermentation. Enzyme immobilization on the other hand is a new technology but is already available, whilst biosensors, biological devices for system monitoring and control, are still being developed.

These observations help to explain why it is anything but simple to define the skills and appropriate training for this new area of activity. The factors to be considered relate to a number of vastly different aspects, ranging from the variety of specialist knowledge required to the particular objectives of different institutions. It should also be remembered that the development of new professional categories is dependent to a far greater extent on practical necessity than on the skills and knowledge acquired during basic training. These facts are important in determining the form and instruments of recruitment and selection which companies should adopt in order to find suitable staff.

## PROFESSIONAL SKILLS ASSOCIATED WITH THE APPLICATION OF BIOLOGICAL TECHNIQUES IN THE VARIOUS SECTORS.

As can be seen from our introduction, it is meaningless to talk of a "typical" biotechnological industrial enterprise or of correspondingl "typical" professional categories. It is more useful to think in terms of a combination of skills which must be attained at various levels and in different situations, in conjunction with, and more particularly instead of, the skills currently prevailing in the various sectors of applied research (Okun, 1984; OTA, 1983; Bevan, Parsons and Pearson, 1987). It is also important to stress that these considerations are valid not only for highly qualified staff (for example engineering, chemical or biological process designers) but also for the auxiliary workforce, such as technical assistants and operatives.

Even allowing for the fact that we have progressed beyond the initial breakthrough stage in the biotechnological areas of recombinant DNA and hybridomas, which are now approaching a degree of technological stability, we must still remember that it is absolutely impossible to speak of a standardized form of work organization. Instead, it is necessary to bear in mind the requirements of specific and often distinctly differentiated processes which must generally be adapted to the products to which they related. It follows that as soon as the traditional production processes are replaced or supplemented by biotechnological processes in the sectors where this is technically possible, various existing skills must similarly be replaced or supplemented by new skills within the companies concerned. Consequently, it is necessary to create a breed of flexible professional personnel. Given the rapid evolution of biological knowledge, these professionals must be willing and able to continue further training throughout their working life, even if only to assist mobility, which in this sector in particular will without doubt remain of immense importance in both the short and medium term.

Currently, graduates (including those with PhDs) and technical personnel with a high level of training in the disciplines of the biosciences are to be found mainly in public and private research laboratories, in particular those of large companies, and in the small businesses generated by university research.

Although both the birth and mortality rate of these small businesses is high, they do have the capacity not only to produce new products or develop new processes of major importance, but also to supply services - ie they possess all the characteristics relevant to the present stage of initial market breakthrough. The types of services supplied range from perfecting and designing highly sophisticated fermentation reactors, through "third party" services such as the provision of advanced scientific data, technical analysis or commercial advice for companies, governments, the media and banks, to the production of scientific instruments or reagents for chemical analysis. Consequently, sales personnel

can also play an important role in the provision of these services, in that users must be taught how to use the new products. For example, biopesticides require methods of use very different from those of conventional pesticides.
Technical staff and graduates with a thorough knowledge of genetic engineering and molecular biology are beginning to be found - although still only rarely - in various large chemical analysis laboratories, particularly in the field of hazardous substances and organisms. The increasing health demands of populations will probably be met by products that are ever more sophisticated even in their use and this trend makes it necessary for diagnostic laboratories to have higher levels of specialization.
(Yoxen, 1987)

According to most of the literature examined in a paper written for the European Foundation (Lanzavecchia, Mazzonis, 1986), there are at present relatively few employment outlets. Even the most favourable forecasts do not anticipate any major quantitative increase in the labour force, stressing instead the changing functions and skills of those already employed.

However, it remains certain that the professionals currently employed in biotechnology-related research by public bodies and large and small private companies will increasingly be joined by other categories in sectors at present outside the high technology professions, in the sphere of agriculture in particular. It is clear that these posts will be filled, at least in the initial stages, by individuals who have already worked with advanced biological technologies in some way, and that pride of place will go to those who have pursued an education sufficiently open and flexible to enable them to adapt rapidly to the particular applications required.

Mobility will mainly be between sectors of the market, but there will also be mobility from the public to the private sector and from small to large companies. This latter point is more than justified by the present economic climate, in particular in the production of therapeutic, diagnostic and nutritional products, where there is a clear trend in Europe, Japan and the USA for smaller companies to be taken over by large combines.

## THE MAIN PROFESSIONAL FIELDS AND ESSENTIAL QUALIFICATIONS

Although it may appear paradoxical, we have found, in order to describe the professional fields involved, no alternative to the conventional classification, which lists three basic spheres for those working in biotechnology and related technologies: research, production and services.

As regards production, a systematic description of the relevant professions and a selection of sample applications can be produceded on the basis of a study by the Centre d'Etudes et de Recherche sur la Qualification (Cossalter, 1986). On the question of public and private research, or research by national or local government agencies, the nature of the work will in general be specialized, so that a definition of professional qualifications will be less clear-cut, in particular in view of the fact that this field is even more in constant evolution than the others. It should however be pointed out that the "new" type of researcher or technician described by some authors is, in fact, very similar to the existing personnel found in research laboratories ever since the post-war years.

In the sphere of services, the professional profile will often be similar to that found in production and research; in sectors such as information, documentation, commerce or consultancy, the capacity for broad generalization will be required. Frequently it will be possible for individuals to move into services from the other two sectors, after developing their specialization to the point where they feel the need to move on and test their abilities in other fields. As regards the technical services of maintenance and control of instruments and machinery, the same considerations apply as for researchers, ie the type of work required in future will be no more than a technically updated version of the present requirement.

Professional specialists, who generally have research doctorates or other higher qualifications, are employed mainly for basic research, technological research and, in companies, research with long-term prospects. This sphere also requires personnel able to help implement technology transfer to the final stages, in other words the development, demonstration and distribution of the new products and processes.

For example, the following are required. Firstly specialists in biosciences are needed to support the development of pure biotechnological research. These scientific personnel will have a high level of specialization in their initial training, and will generally have further international experience in various basic disciplines such as genetics. Secondly bio-engineers in the general sense are needed. These people will be united by a common disciplinary denominator and a basic biological competence, with a mixed background in biology and chemistry supplemented by a knowledge of engineering science (in particular chemical engineering suitable for working with living matter), and they must also have basic grounding in disciplines such as biochemistry, microbiology, immunology, genetic engineering, enzymatic engineering and purification technologies and biophysical chemistry. At the same time, they must not be totally unfamiliar with disciplines such as applied mathematics, data processing, organic and analytical chemistry, thermodynamics and industrial engineering. Finally, all bio-engineers, in particular those working in company laboratories, must take account of the problems encountered in the transfer and adaptation of scientific knowledge to various industrial sectors, must be acquainted with economic and social sciences and should have some knowledge of the problems of intellectual property rights, labour organization and health and safety.

In different countries these individuals will be educated in different environments, but primarily in the specialized engineering colleges which in the last few years have started to bring together the disparate biological information available from the various specializations found in standard university courses and postgraduate fellowships. As in the other professions, there is a fundamental difference between the research carried out in small and large companies: whereas large companies are generally interested in specialized personnel in a strictly defined field, smaller companies require not only researchers as"authors" of new products - ie the actual protagonists of business in the high-tech industries - but also technically qualified experts with solid backgrounds in economics, marketing and law.

These "generalists" must in fact be able to evaluate the multiple aspects of a biotechnological project and to propose solutions to problems of technology transfer, prototype development and industrial scale-up of biotechnological products and processes. (Colombo, 1984) They must also be able to propose or justify an industrial biotechnology option in relation to other possible technological options, and, if necessary, to re-adapt with no major problems to the other sectors of activity.

In companies supplying services to third parties, laboratory technicians are also important, to be trained concurrently with the development of productive processes, transferring their knowledge to work with the new biological methods.

Although there are differences, the professionals required in services, including support services, can be compared to those involved in business and research. All diagnostics- laboratory analysts using advanced biotechnology, or all technical workers carrying out chromatographic column separation, will be working in the same way, in the same time-spans and with the same organizational structure, whether they are working in industrial companies or in service laboratories. The situation is different in the case of more specialized services: technology transfer experts or companies providing consultancy services to companies, banks or governments must have a greater capacity to work horizontally - in other words, they must supplement their professional qualifications with the talents of the "generalist".

## OTHER PROFESSIONAL CATEGORIES AND GENERAL CONSIDERATIONS

With the development of activities in the biotechnology sector, careers and professions are constantly enriched by a variety of new facets. For example, there is the use of specialized architects for the detailed design of laboratories (Shaffer, 1983) and for the biotechnological infrastructure. In addition, the potential hazards to employees, populations and the environment posed by the special materials used in advanced biotechnology make it essential to create a body of professionals specialized in the legal and insurance problems of laboratory and plant mangement and organization (Webber, 1984). Similar specializations will be required of the public servants entrusted with updating the requisite legislative apparatus and inspection structures. The legal professions will also be responsible for patents (Davis, 1984; McGarity, 1984).

The World Health Organization stresses that, in view of the human implications of the use of biotechnology, it is necessary to create specific specialists in industrial health and safety in order to monitor these operations in the various countries, including developing countries (Miller, 1983; Flagg and Purnell, 1983).

In designing plant, great attention must be paid to the related human and environmental risks: the requisite knowledge must therfore be incorporated in the qualifications of technical design staff, in exactly the same way as in other sectors (the nuclear industry is a good comparison).

Other specialistss increasingly in demand include experts in instrumentation, both users and repairers. Bruni Kobbe cites a company in California which produces protein synthesizers and which has been increasing its workforce by approximately 400 people per year, mainly experts in electonics and computer processing with a knowledge of biochemistry and chemical analysis. (Kobbe, 1986) This example is typical of the trend in many sectors, whereby highly specialized personnel are created by the convergence of different disciplines. Starting from a common biotechnological base, these specialists combine the knowledge and use of automation, electronics, sensors, lasers and new materials: in biotechnological processes, the requisite product purification can and must be controlled automatically, whilst columns can be of varying composition and sensors can vary according to the nature of the process. The same arguments can be applied to other disciplines, such as electro-engineering, and to other professional categories, such as specialized technicians.

In addition, this type of professional is already present in the pharmaceutical industry, although certain differences exist, and is also quite common in biotechnology companies which use a specialized workforce in this particular field,, drawing employees for both production and R & D from the chemical sector.

In this way, there is a form of competition between the typical small to medium-sized biotechnology companies and the large pharmaceutical groups: the smaller enterprises offer the qualitative attraction of high flexibility and creative freedom, whilst the larger concerns offer the bait of higher salaries, greater job security and often, in particular for researchers, better technical facilities at their disposal.

## TRAINING

There is no single specific training for careers in biotechnology and bioengineering. These areas are multidisciplinary, born of the combination of genetic engineering, microbiology, enzymology, biophysics, biochemistry, organic chemistry, chemical engineering and molecular biology. Chemists and biologists can become biotechnologists. Biotechnology specialists can also come from higher agricultural colleges, universities with agricultural courses, engineering colleges (including in France the Grandes Ecoles) or, finally, from recently founded universities like the Technological University of Compiegne.

Current estimates show that today, in the USA and Japan, one third of personnel manipulating genes are molecular biologists and immunologists. Large numbers of molecular biologists specialize in human and animal molecular biology, because this type of research has numerous applications in medicine and is heavily sponsored by public institutions, which in recent years have recruited many medical and biological researchers. However, there is currently a shift towards plant breeding, in view of the boom in biotechnological applications in agriculture and the degree of public interest and private finance in this area.

Many of these specializations are obtained by postgraduate courses and diplomas (including doctorates), except perhaps for biochemical engineering. Qualifications are obtained from the universities, without further specialization, and also, obviously, from the French-type Grandes Ecoles.

Engineering, without further specialization, can now provide industry with personnel able to work on a large scale with reactors, including biological reactors. For biotechnologists - and consequently engineers - to obtain a fuller professional training, their preparation must touch on four different areas: theory, application, economics and practical experience in collaboration between universities and industry. In addition, the economic side of their training must enable these technical experts to evaluate the costs of the various competing processes in national and international industry.

The problem of recruiting qualified personnel is at present still a question of quality rather than quantity: ready-trained individuals are not always available from amongst those who have completed their basic professional training, and they must be specifically trained or in some way "retrained" to meet the requirements of each company. In the short-term, there is no easy remedy to this problem because, as for electronics and automation, a whole series of trial-and-error evaluations are still required in order to define profitable avenues of research. This process will require time, given that it will follow the rate of maturity of the sector.

Despite this fact, some elements of horizontal specialization already exist at the high level of qualifications provided by certain research doctorates for "genetic engineers". Some training programmes already combine the various biological disciplines - molecular and cellular biology, biochemistry, genetics, etc. - with knowledge and techniques drawn from pure chemistry and physics.

To ensure effective biotechnological training systems it is necessary to take account of the variety of knowledge required, the variation within each specific profession, the predominance of one disciplinary field over another, the fundamental need to master the essence of methodologies and techniques rather than to learn particular crafts, and the trend towards flexibility of application rather than strict observance of ready-made protocols and procedures.

Training must also take account of the fact that, as we have seen, in each sector of application it is necessary to assure the functions of research, design, development and technical and operational support, together with functions of a more purely managerial and strategic nature. The challenge presented by this type of combination lies in linking these different aims in order to provide good communication between the various professional and technical categories, by ensuring that both those closer to research and those oriented towards management (even if of a new type) are supplied with complementary knowledge drawn from their respective fields of specialization.

In other words, the basic specializations on which these professions are founded are certain to be vigorously reinforced. Their purpose and capacity will also be extended according to the identification of individual work objectives and the constant striving to achieve them in practice.

Particular stress must also be laid on the definable management support professions, which must have a high capacity to combine information and to take decisions: an ability to see things as a whole is needed to identify more clearly the progress of the production processes in which there are many variables in constant flux, whilst the capacity to take strategic decisions is needed in an industry that is evolving rapidly, without a tendency towards routine series production.It is also necessary to control the risks of over-specialization: biotechnology is made up of various interrelated disciplines, and its practitioners must be able to work in interdisciplinary groups.

At present, this process is already taking place in industries applying biotechnological disciplines: chemists, microbiologists and process engineers are constantly employed in interdisciplinary working groups, with all members providing their own specific skills in constant interaction with the others.

Advanced training must also be organized on various levels, in parallel with the professional categories described above, and must aim to expand the knowledge of a) process operators, who at present are trained mainly by some form of on-the-job training system; b) specialized biologists and biochemists, who at present normally have a postgraduate qualification; c) engineers employed in the design and control of biotechnology plant; d) management, currently lacking the specific knowledge required for useful employment in this type of company; in this group, we must not forget apparently secondary professions such as brokers and experts in marketing, legal affairs and insurance, etc.

In addition, even allowing for the traditional diversity of training methodologies for all levels, a biotechnological "culture" must be transmitted to the entire workforce employed by biotechnology companies, as has already occurred in the world of computers.

Besides education prior to employment and on-the-job training for personnel of high and medium-level specialization, attention must also be paid to the training of lower-level staff employed in maintenance, monitoring and cleaning for biotechnological production (Dimond, 1985).

This lower-level training is important not only because the work involves handling living organisms which may occasionally have pathogenic effects, but also because the delicacy of the materials processed can require a greater degree of participation. It is essential that young women intending to have children and other high-risk sections of the population should be informed in detail of the products with which they come into contact.

These conclusions indicate the need for contiued training for all, in order to demonstrate that, subject to the obvious environmental criteria, when the advanced industrial applications of biotechnology are diffused and adopted, they can help give the workplace a common culture, making work more creative. By contrast, the lack of training at all levels can nurture the alienation and fear of the workforce when faced with new processes, thereby enormously limiting not only their capacity but also their pride in their work.

# 10 Human Resource Planning in Biotechnology

Richard Pearson

## INTRODUCTION

Biotechnology is, along with microelectronics, information technology and new materials, one of the key 'new' technologies which will radically reshape the industrial world and employment in the late twentieth and early twenty first centuries. In the late 1970s and early 1980s, biotechnology became the glamour technology, attracting many millions of dollars from venture capitalists and governments alike as they sought to capitalise on the latest advances in biological science and genetic engineering.

The aim of this paper is to review what is known about employment in biotechnology, its skill needs, how these needs might develop in the future, and the implications for the labour market and for the education and training system. The paper focuses on the UK, US and Europe. It draws comparisons from other new technologies where appropriate. But first what is biotechnology? It is a science capable of many differing definitions, because it builds on a range of disciplines. For the purposes of this paper biotechnology is defined as 'The application of biological organisms, systems and processes to manufacturing and service industries'.

It is clear that biotechnology is not focussed on a single industry, activity or group of products; rather it is better considered as a process technology with implications for such diverse sectors as agriculture, environmental control, fuels, healthcare and pharmaceuticals. One of the key characteristics of biotechnology is its inter-disciplinary nature, embracing such diverse skills as biochemistry, biology, microbiology, genetics, medical and veterinary sciences and bioprocess engineering. It is still, however, an emerging technology and one for which the boundaries, priorities and skill needs are changing.

Biotechnology has a number of features that have to be borne in mind when considering key skills. First is the distinction between the traditional biotechnology activities of brewing and food, and novel biotechnology, as the available evidence suggests that the transferability of skills between the sectors is limited. The focus of this paper is on novel biotechnology. A distinction has also to be drawn between those non-specialists employed in biotechnology support activities including the highly qualified computer scientist, the support technicians, and the administrators; and those whose primary activities and skills focus on biotechnology. The latter group is the one of prime interest here as they have relatively dedicated and non-transferable skills and the lead times for their training and development are relatively long. Third, in terms of occupations there is no such thing as a biotechnologist. The interdisciplinary and applied nature of biotechnology means that a professional worker in biotechnology is better defined in terms of their work or research experience, such as fermentation technologist or molecular biologist, reserving the term biotechnologist as a very broad generic title. Finally the inter-disciplinary nature of biotechnology and its dispersal across industrial and academic boundaries means that assessments of manpower and training issues can be complex. As such definitions and boundaries have to be drawn on pragmatic lines and care has to be taken when comparing and referencing different sources.

## EMPLOYMENT

The first and broadest international assessment of the rapidly developing biotechnology labour market was that undertaken by the Office of Technology Assessment of the US Congress (OTA, 1984). It estimated that there were, in 1983, about 5,000 people employed by companies in biotechnology research and development in the United States. There were however no estimates for either academia or the public laboratories. Although the OTA reviewed developments in biotechnology across a number of countries, concluding that the main activity was to be found in the US, Germany, Switzerland, France and the UK, with Japan at that time lagging behind these countries in its developments, it did not find any estimates for overall employments in these countries. While generalised estimates of the position in individual European countries have been made attention to the question of availability of key skills has been limited. (Lanzavecchia and Mazzonis, 1986) The most detailed national study has been that of the UK. The first estimate for employment in the UK was that there were about 2,000 employed at professional level in the commercial sector, academia and the public laboratories, with about half the total being employed in the commercial sector. (Parsons and Pearson, 1983) Using some gross assumptions it has been estimated that total worldwide employment at that time might have been under 30,000, a smaller figure than for those working at professional level in the semi-conductor industry, perhaps the most closely analogous sector. (Pearson, 1984) By way of contrast, the total employment of IT professionals world wide runs into several millions. It can be seen that in the mid-1980s at least that biotechnology is not a numerically significant employer by any criterion.

## THE EMPLOYERS

Commercial organisations are not the only ones active in biotechnology, there are also publicly funded research institutes, and the universities and other academic institutions. Overall in 1986 it was estimated that in the UK there were 100 industrial organisations, 20 publicly funded research centres, and some 120 departments in 53 universities, polytechnics and medical schools active in novel biotechnology.

In the case of industrial sector this represented a doubling of the number of organisation since 1983. They embraced a wide range of activities including R&D, production, plant engineering contracts and industrial service organisations. The main activities were, however, still R&D, with only a few companies involved in large scale production. The companies were also spread across a wide range of sizes, from small recent start-ups with five or fewer professional staff to divisions of major multinationals, several of whom employed 100 or more professional biotechnology staff at graduate level and above. Employment in the commercial sector has grown by about a third in the last 3 years to total about 1600 at gradute level and above.

Among the research centres reorganisations have led to a slight contraction in their numbers although employment in novel activities has expanded by three quarters in the last 3 years to total about 1000 at graduate level and above in 1986. Employment is, however, far more concentrated with some 20 centres active in 1986, five of whom each had 100 people engaged in novel biotechnology activities.

Higher education plays a vital dual role embracing both a research and training function, in the case of biotechnology the latter being an integral part of the former. The vast majority of specialist skills are concentrated in the research area as to date training (outside of short course provision) concentrates on the basic scientific disciplines of biochemistry, microbiology etc. and does not require many staff specialising only in biotechnology skills. Separating out the biotechnology specialists here is rather harder, but they were estimated to total about 700 at postdactoral and higher levels.

Although the employment characteristics of each of these sectors is different the increasingly close working relationships between them is making the boundaries increasingly blurred as, in particular, higher education sets up its own start up companies and consultancies, and research institutes and some industrial companies play and increasing role in training at postgraduate level.

In the United States the number of companies active in R&D has more than doubled over the last three years and totalled some 500 in 1986. The only available estimates show that employment in industrial R&D companies has grown at a faster rate than for the UK, up from 5000 in 1984 to 7000 in 1985 and on to 8000 in January 1986.(NSF, 1987) Parallel estimates are not available for Europe.

## KEY SKILLS

What then do we mean by a professional biotechnology worker? We have already seen that there is a wide range of disciplines involved. In terms of educational level then employer-based studies show that, outside of bioprocess engineering, the qualification profile is very high, with workers rarely being regarded as a specialist professional until they have had several years post-doctoral experience. In bioprocess engineering related activities then the qualification threshold was generally lower, at first or masters degree level. This reflects in part a more general distinction between pure science, where doctoral and sometimes post-doctoral, qualifications are often essential for long term career progression, and engineering and technology, where such high level, academic-oriented qualifications are rarely sought outside academia.

The scientific or technological specification of particular posts, or even of whole R&D teams or activities, is generally highly specialised. It is not therefore appropriate to define key skills in a conventional occupational or academic sense but rather in terms of principal generic skill inputs which involve: (i) genetic manipulation, where particular skills of importance are immunology and molecular biology, and plant molecular biology; (ii) fermentation process technology to develop the scale up of production processes; and (iii) linked specialities embracing the above two areas including enzymology and cell culture.

These represent the biotechnology-specific skills, in addition to which there are the adjacent specialists in medical/veterinary science, chemistry and computer science, along with laboratory support and maintenance skills.

If this is the current range of high level skills, what is the balance in terms of employment? In the US in the commercial sector one in three staff is said to be trained in genetics or molecular biology. Another third were scale up plant and downstream processing specialists in bioprocessing engineering or microbiology and biochemistry. (NSF, forthcoming) In the UK the sharpest growth in demand has been for skills in genetic engineering, it is now the largest skill group in higher education (40 per cent), and research centres (39 per cent) with a rather lower density in industry (26 per cent) where the major skill group is in bioprocess technology (40 per cent), although there were major differences between individual companies (Bevan et al., 1987).

## SKILL SHORTAGES

Skill shortages in biotechnology are selective rather than widespread, to some extent this should not be surprising given the relatively small numbers involved and the broad range of base disciplines on which it draws. The most recent evidence on the UK shows that recruitment difficulties for industry and research centres were largely confined to highly specialised posts where the quality rather than the quantity of the applicants was the problem. (Bevan et al., 1987) Generally these employers were competitive in the market offering attractive salaries and permanent posts. Although the skill shortages were not widespread the range of disciplines is widening and now embraces plant molecular biology; culture specialists; microbial physiologists; and specialists in downstream processing. Recruitment difficulties were more severe in higher education, where short term contracts, inflexible salary structures, and poor career prospects and facilities all combined to make it extremely difficult to attract able PhDs to post doctoral research and teaching posts.

In the United States there has been seen to be an abundance of personnel trained in basic biological sciences and the ready availability of molecular biologists and immunologists was seen as one contributing factor to the rush of small start-up companies and the initial American lead in biotechnology. While there was not major concern about skill shortages in the mid 1980s, industry is acutely concerned that there may be a shortage of well trained bioprocess engineers in the future which will constrain the design, development and monitoring of biological scale-up production. The other area of potential concern was that of industrial microbiology where demand is increasing rapidly but supply from higher education remains static. While industrial and research centres were not reporting problems, concern in higher education about skill shortages was rising under the combined problems of an outflow to industry and a shortage of students staying on for post doctoral research and seeking faculty positions.

Evidence about shortages elsewhere is more limited. In Japan there has been a abundant supply of bioprocess engineers and industrial microbiologists but there have been shortages of experts in genetic manipulation and basic molecular biology. Japanese companies have, however, started intensive training programmes, in Japan and overseas, especially in the United States where Japanese scientists have a major presence at the National Institutes of Health and they also pay for training placements with commercial companies. They are also seeking to encourage job mobility between sectors. A recent report by SERC suggests that these policies are being successful and that Japan is rapidly catching up in terms of its basic molecular biology skills. (SERC, 1987)

Within Europe the limited evidence available suggests that Germany has a shortage of molecular biologists but that the supply of bioprocess engineers and industrial microbiologists is adequate. France appears to have the most serious shortages and there are reputed to be few specialists in many of the areas of biotechnology and only a limited number of internationally recongnised reserach centres. Major initiatives are being undertaken to remedy those deficiencies but they are not expected to solve the short term problems and international recruitment is likely to remain a priority. Italy apparently has a relatively undeveloped biotechnology base and only limited skills while other European countries also have smaller levels of activity although in the case of Denmark, Belgium and Ireland there are specific local strengths. (Lanzavecchia and Mazzonis, 1986)

## BRAIN DRAIN

One consequence of skill shortages is that market forces come into play to force a redistribution of scarce skills towards the most attractive employers, who may be appealing not just in terms of traditional factors such as salaries but in research, as indicated by levels of support staff, facilities, research freedom and working environment. The last decade has seen these adjustments follow two important dimensions, across international boundaries, and from higher education to industry. If these flows remain one way they will seriously undermine individual countries' long term ability to develop their biotechnology base.

The United States has a long history of immigration to meet its need for key skills and concern has existed for many years as to the effect the brain drain was having on Europe and the rest of the world. The rush of money into biotechnology in the late 1970s, with start up companies offering very high salaries and benefits packages including share ownership schemes was one very visible contribution to the flow of staff into the United States. The brain drain, and its apparent scale, causes and consequences, has been the focus of a great deal of debate in the UK in recent years and there has been one major study of the biotechnology brain drain. (Pearson and Parsons, 1984) This estimated that in the decade to 1984 about 250 biotechnologists had gone overseas representing in aggregate about 13 per cent of the stock in 1984, although a smaller, but unquantified number had also returned to the UK over this period. The peak flow was in 1981 and 1982 and this has since abated. The main destinations were out of Europe and included the United States (46 per cent) followed by Switzerland (16 per cent) then Canada, Sweden and Australia. The largest loss was from UK higher education, the main recipients were commercial organisations. More than 65 different overseas orgaisations were known to have recruited UK nationals over this period, one major start up in Switzerland taking at least 16 senior scientists in a two year period. Typically the leavers have been young post docs, although in the early 1980s there was a significant number of senior staff leaving. Most went to improve career prospects and working conditions and not for salary reasons. Few expected to return as they saw little prospect of the relative position of UK science improving and they were then also aware that a return would entail a cut in their salary levels. While the outflow did not have significant consequences for individual employers, and numerically it was not a significant cause of shortages in the UK, the flow overseas did represent a loss of (possibly under-utilised) talent and, perhaps more importantly, was seen as a symptom of the poor conditions prevailing in the higher education. Another manifestation of the latter has been the reported flow from jobs, and declining interest in post doctoral fellowships, in higher education, to posts in industry or research centres who are able to offer better salaries and working conditions. This latter flow is not confined to the UK but is also seen as a serious problem in the US (OTA, 1984) while both flows are of concern in Europe where overseas losses to the United States were a particular concern.(Lanzavecchia and Mazzonis, 1986)

## FUTURE TRENDS

The last decade has seen a significant growth in employment levels of professional staff in biotechnology, although in the last couple of years the 'industry' has gone through a phase of consolidation as the euphoric promises of many of the new start up companies have failed to materialise. Many of these have closed or contracted, or been taken over by the established multinationals, the most recent development being the takeover, by Glaxo, of Biogen's much celebrated but now much reduced Swiss research centre. The real development of biotechnology is clearly going to take into the next century and there are likely to be many more phases of adjustment before it reaches full scale maturity.

Although significant growth in the employment of specialists is expected along with perhaps rather higher rates of growth among support staff as the technology moves into large scale production, the size of the activity is such that it is not expected to have massive direct job creation potential. Indeed the very nature of the technology is that in many areas it is one of displacement, and jobs are expected to be lost in many traditional areas such as agriculture and environmental control.

Forecasting future demand in an uncertain world is not possible with any precision. However, the available evidence points to continuing growth in the demand for professional staff in biotechnology. For example, in the UK the majority of commercial orgainsations expect their employment levels in biotechnology to increase in the next five years although the numbers involved remain relatively small. In the case of research centres and higher education then levels of public funding are the key, but most organisations expect demand to continue to grow significantly, especially with respect to short term contract posts. Overall demand could rise by as much as 10- 15 per cent per annum with growth being driven by the expansion in higher education. This represents a continuation of recent growth trends. The balance of skills sought is not expected to change significantly although there will be a growing emphasis on plant molecular biology, in downstream processing, protein engineering, and biochemical engineering.

Biotechnology draws heavily on its base academic disciplines and higher education, primarily at postgraduate and postdoc level, is likely to remain the key source of new skills in the future. This is in contrast to other new technologies such as microelectronics and information technology where higher education provides the basic skills, but the higher level skills are normally developed in industry.

One response of higher education in the UK to meet the need for more specialists has been the development of one year Masters degrees, and indeed a number of first degree courses in 'biotechnology'. This has happened despite the near unanimous opinion of employers that students qualifying in this way were of little or no relevance to their manpower needs. Indeed among the Masters courses now running graduates from only one, specialising in biochemical engineering are able to move into relevant biotechnology employment. (Bevan et al., 1987) A more successful development however has been the provision, by higher education, of short courses for industrial and research organisations. Over 52 were planned to run in 1987, a massive increase on the seven that ran in 1983. (Parsons, 1987)

If future expansion in demand continues at the expected rate then selective skill shortages, especially for experienced people in plant molecular biology and in key specialisms within bioprocess technology are likely to intensify, although the main burden is particularly likely to focus on higher education. In the longer term one major supply constraint could be the lack of growth, and possible decline in the throughput of PhD's and post docs in higher education who provide the base supply of skills for biotechnology. This possible decline will be driven by a number of factors. First the significant demographic downturn over the last decade which will reduce traditional student demand for places in higher education in most European countries and the US; second intense competition by industry for good graduates will discourage those that do qualify from staying on to postgraduate degrees and training, where the rewards and facilities are often poor; and third the likely continuing shortages of faculty and resources to support such training. (Bevan et al., 1987) Such problems are already manifest in the UK, Europe and the US, and are expected to intensify in the future.(OTA, 1984; Pearson, 1984; Lanzavecchia and Mazzonis, 1986)

There is a delicate balance between manpower supply and demand in novel biotechnology, with key staff operating in an international labour market. Its development depends critically on the availability of key skills. The technology is not expected to be of significance in terms of job creation and may indeed lead to the displacement of many jobs as applications develop towards the end of the century. The specialist skills required to support its development are at a very high academic level and could be increasingly in short supply unless the level of investment in the subject in higher education is improved and particularly the throughput of post doctoral students in key disciplines is increased in the future. To overcome the expected growing shortfall of key research skills in the United States in the 1990s the NSF has forecast that there will have to be an increased emphasis on overseas recruitment, which implies a further resurgance in the brain drain from Europe.

The likely trends in the balance between supply and demand for key skills for biotechnology within each European country is unclear. Indeed the current level and distribution of skills is not known. New initiatives are being undertaken to develop closer linkages between higher education and industry including trans Eurpoean initiatives such as COMET. Problems are, however, likely to remain in the future and to be in highly specialised areas involving relatively small numbers of people. Further research should focus on a more detailed analysis of the position in individual countries, the ways in which skill needs are developing, whether increasing specialisation is likely to remain, on the effectiveness and relevance of new types of degrees, of higher education/industry linkages, and of national initiatives aimed at improving the supply of key skills. Overall there need to be a continuing monitoring of the trends in demand and a closer analysis of trends and key influences on the changing supply at postgraduate and postdoctoral level within indivdual European countries.

# PART FIVE: PARTICIPATION

# 11 Public Debate on Biotechnology: The Case of Denmark

Jorgen Lindgaard Pedersen

## INTRODUCTION

During the ten year period 1976 - 1986, Denmark has changed its approach towards biotechnology from a non-policy to a public regulation of the different applications of this new technology in society. My purpose with this paper is threefold: firstly, to describe the main features of this public regulation in order to illustrate how it came into being; secondly, to give some of the recommendations from a conference held by the Danish Technology Board in cooperation with the Society of Danish Biologists in April 1987; thirdly, to make an evaluation of the Danish public debate and policy-making in order to extract some lessons of possible interest to other countries.

## REGULATION OF SAFETY AND ENVIRONMENTAL MATTERS AND R&D-POLICY

In the following, the use of the term biotechnology will in fact refer to just one important technology in the whole biotechnological field, namely genetic engineering or recombinant DNA technology. Genetic engineering is used commercially in production processes using microorganisms in closed fermentation systems, in the release of microorganisms in the open field, in plant breeding, in animal breeding. As Fresco's paper in this volume shows it is also being applied in human genetics, with the eventual intention of altering genes in human beings.

In 1986 the Danish Parliament, the Folketing, passed a new law regulating all applications of genetic engineering in research and production, with the exception of the application of genetic engineering in human genetics. The purpose of the law was to protect working conditions, the environment, the health of the general population and nutritional quality and food safety, where genetic engineering was employed. Its passage ended a lengthy discussion among many types of experts, science journalists, parliamentarians and the general public about the need for a special law and administrative apparatus in this field. Nearly all experts from business and the scientific community were from the very beginning against a special law. The parliamentarians took the view that regulations were a necessity, especially in the area of human genetics. The environmental groups and most of the science journalists believed that regulation was the best way for society both to control and to understand the advances made in this technology.

Through the efforts of a few members of parliament from the opposition (the Social Democratic Party and the two Socialist parties) and members of the Social Liberal Party outside the government, inspired by science journalists and the environmentalists (and possibly furthered by some internal disagreement in the coalition government parties), it proved possible to pass a law in the summer of 1986. Since then, the Minister of Education has appointed a so-called Gene Technological Council with special responsibility to consider possibilities and problems in relation to research. This degree of statutory control is widely seen as significantly more strict than the comparable set of regulations in force in any other Member State of the EEC. This makes both the genesis of the law and its application in practice particularly interesting to many observers.

The 1986 law has as its central principle, that it is forbidden to import, utilise in production, or release in nature gene-spliced microorganisms or cells, unless this has been permitted by the relevant authorities. The intention of the law is that the applicant can expect to obtain permission to use microorganisms in industrial production if the safety measures are fulfilled. In order to obtain permission, the firm or laboratory must make - at its own expense - an analysis and an assessment of environmental and health risks to the workers in the firm or laboratory and to its neighbours. The important questions to be answered are: Is the organism able to survive in the environment or in nature? Is the organism able to reproduce itself? Can the organism spread out and establish itself in the environment or in nature outside the site of leakage, discharge or exposure in nature? Is the organism able to transmit its genetic material to other organisms, so that its genes become established in wild-type populations, even if the original organisms themselves die ? Finally, an assessment must be made as to whether or not the organism or genetic material from it can harm or have other unwanted effects on health, the environment or nature.

With regard to the release into the environment of microorganisms or plant materials treated by genetic engineering, more restrictive safety standards can be expected to apply than in the case of microorganisms in closed fermentation systems. Thus far in Denmark, no approvals have been given to such projects. Finally, the law establishes that the decisions made by the authorities approving production and release shall be made in the form of public notifications and in written form to organisations and individuals who clearly have an interest in receiving information about approval from the authorities, and who are entitled to appeal in relevant instances. A fundamental premise of the 1986 law is the contentious assumption that it is not possible today, on a scientific basis, to predict the behaviour of a gene-spliced organism, after it has been released. Since it is not possible in advance to deny the possibility of unwanted effects in the application of genetic engineering, the law requires an assessment, case by case. This is the same principle as recommended by the US Environmental Protection Agency (EPA). Such risk assessment will, at the same time, serve as a repository of experiences, which can provide the basis for a later, more general, regulation and better preparedness in the case of accidents.

The Danish law is administered by a special office within the Ministry of Environment, the Food Directorate. In the last phase of the political debate, especially from 1983, this Ministry became actively engaged in the efforts to get a special law passed. During the ten-year period prior to the passage of this statute in 1986, there was a growing awareness of the desirability of devising a special law, or at least a series of regulations, governing the industrial application of genetic engineering.

The story begins with an initiative taken by the Research Councils (the primary external source of academic funding in Denmark) in 1976, when they created the so-called Registration Committee. Members of this committee included representatives from the Research Councils, industry, the Royal Danish Society of Sciences and Letters, the Ministry of Environment and the Danish Health Directorate. The committee's task was to establish a register of all experiments in genetic engineering in Denmark. In addition, the committee was to make proposals for public prescriptions in the future, if it found this to be necessary. The registration, it should be stressed, was voluntary. In the following ten years, the committee followed the American guidelines from the National Institutes of Health (the so-called NIH-guidlines). Recently, it has also drawn on recommendations and reports from international organisations such as the OECD, the European Commission and the European Science Foundation's Liason Committee on Recombinant DNA Research.

The thinking of the Registration Committee can be summarised as follows. There has been a growing consensus in the international scientific community, that genetic engineering under laboratory conditions does not carry risks beyond those that always exist in microbiological laboratory experiments. The same is true for biotechnological production on an industrial scale. For this reason, there is no need for a special law in Denmark. The legal basis for all relevant regulation could be found in the existing laws, especially in the laws regulating working conditions. Nevertheless, it was felt that all experiments with genetic engineering should be registered, perhaps by law.

In 1986, the chairman of the Registration Committee added that the creation of a special law could lead to exaggerated bureaucratization with unfavourable consequences for the competitiveness of Danish firms. But he accepted the possibility that it may become politically necessary to create a special law. The committee was abolished by the end of 1986. All in all, it had registered about 130 experiments. Only four of these were experiments with direct relevance to production. Of all the experiments registered, about 10 per cent were placed in category 2 of the NIH Guidelines, the rest in category 1. It should be noted as an important feature in evaluating the effectiveness of such a committee, that one of the most important industrial experiments, in fact the first Danish use of gene-spliced microorganisms in industrial production, was not registered by the committee. The firm explained that they did not want to give information to their competitor, one of whose employees at on the committee. The initiative to draft a law was taken in 1983 by the then Minister of Interior, who desired to bring this area under ministerial control. She wished, moreover, to give the Health Directorate, which has a quasi-independent status vis-à-vis the Ministry of Interior, control over all experiments in genetic engineering. The Minister of Interior recommended liberal regulation of industrial applications of genetic engineering in order to give both industry and agriculture the best conceivable working conditions. During the preparation of the law, the initiative passed from the Minister of Interior (a member of the Liberal party in the four-party Liberal-Conservative minority government) to the Minister of Environment (a member of the Christian party in the government). The result was the much more restrictive law as described above.

There were two major reasons for this shift. First, there was a change in attitudes among the civil servants in the Ministry of Environment regarding the relevant public interests to be safeguarded by law. Second, there was a strong growing political interest in the whole field of biotechnology, particularly on the part of several highly influential members of the Folketing.

Both the civil servants in the Ministry of Environment and the parliamentarians were assisted in their work by a broad range of technology assessment sources. One important source was the Danish quarterly "Naturkampen", a socialist magazine that deals with science, technology and medicine, which in 1981 published an issue devoted to information technology and biotechnology. The articles were relatively positive towards these technologies, viewing them as potential solutions to pollution and other problems. Another source used by some of the parliamentarians and civil servants was the technology assessment project, PEGASUS, financed by official technology assessment funds administered by the Council of Technology. Finally, the civil servants in the Ministry of Environment, in cooperation with their Minister, were in close communication with the important members of Parliament in the relevant Parliamentary committee.

## HUMAN GENETICS

In the field of human genetics, the debate in Denmark has been no less intense than it has been for industrial and agricultural applications of genetic engineering. But in the beginning of 1987, the Folketing, following a proposal from the Minister of Interior appointed an Ethical Board. The purpose of this board is two-fold: first, to formulate rules for medical activities in this field and second, to advise the Folketing and the government in ethical matters. The proposal prohibits all experiments with the exception of in vitro fertilization in individual cases until the Board has made rules for this area. The proposed Ethical Board will, like the Research Councils, have a certain degree of independence from the Minister, and a closer connection to the Folketing than is usual in the Danish Parliamentary tradition.

The proposed Board is also required to have a substantial number of non-experts among its members. This condition reflects the fact that the parliamentarians have been interested in human genetics for a fairly long time, since 1980 to be precise. Published statistics on in vitro fertilization experiments at the Central State Hospital in Copenhagen inspired a prominent liberal politician in that year to initiate political actions in both the Folketing and the Council of Europe. A hearing arranged by the Council of Europe in

Copenhagen in 1981 increased parliamentarians' interest in the subject. Certain members of the Parliament were provoked by what they saw as arrogance on the part of some of the participating experts. This interest led to several questions in the Folketing directed to the Minister of Interior concerning his (and later her) attitude towards known and possible experiments in this field. This was followed by a public debate regarding the need for rules or legal activities. The members of parliament referred in this debate to articles and discussions by science journalists in magazines, newpapers and the electronic media. Subsequently, in 1983, the Ministry of Interior organized a public hearing and appointed a committee, whose task was to publish a report on the ethical problems raised by genetic engineering, egg transplantation, in vitro fertilization and prenatal diagnostic techniques.

While this work was in progress, a minority in the Folketing proposed on January 1st 1982 a two-year moratorium on all experiments using these technologies outside their original fields of application. The proposal was not passed, but undoubtedly it led to an intensification of the work in the Ministry of Interior. The report from the Ministry of Interior "The Price of Progress" was made public in October 1984. It was commented on by a broad group of organisations during the following half year. In March 1985, the Minister of Interior submitted a review statement to the Folketing, "Ethics and Medical Technology". After a further half year of debate, the Minister introduced to the Folketing a proposal for a law which, after a debate with the opposition, took on the character described above. Two important ongoing themes in the debate between the government and the opposition have been, first, how to secure a reasonable balance between use of expert knowledge and use of lay experience, and second, how to maintain political responsibility where both the Minister and the Folketing were involved. The beginning of the debate about human genetics in Denmark coincided with the first implementation of the recommendations of the Helsinki Declaration, passed in 1975. This declaration deals with experimental medical research on human beings. Seven regional Scientific Ethical Committees and one central committee were appointed. The committees set up on the basis of civil law, were given equal representation of biomedical researchers and laypeople. Their task was to register and approve all medical experiments on human beings. But they did not have formal sanctions at their disposal.

To most members of Parliament, the Scientific Ethical Committees are not suited to solve problems related to gene therapy and gentic engineering. For one thing, the committees are appointed on a civil law basis. For another, the committees must explicitly take as their starting point already existing individuals. As a result, a good part of the gene-splicing activities are excluded from the committtees' competence. Finally, the parliamentarians find it important to vest the moral-ethical responsibility in the Folketing, though without completely abolishing Ministerial responsibility.

## MEASURES PROMOTING RESEARCH AND DEVELOPMENT IN BIOTECHNOLOGY

As the first wave of concern regarding genetic engineering disappeared in the late 1970s, university scientists and civil servants in different parts of the public sector began to make proposals for particular subjects in research and development in biotechnology, which were held to be especially worthy of state support. There were four important initiatives in the public sector.

The first was the Initiative Committee concerning Gene- Splicing appointed in May 1981 by the Danish Council of Technology, which is under the Ministry of Industry, but has a quasi-independent status with representatives from industry and the trade unions and civil servants. The first initiative from the Committee was a proposal to start a gene-splicing group which should be financed partly by funds from the Danish Council of Technology, partly by industry and other with more specific interests. The proposal was accepted by the Danish Council of Technology and the group was established in October 1982. In 1984, the Committee published a plan for activities related to genetic engineering. Its focus was on efforts to create more effective microorganisms. The programme was accepted but it was not a success. The Committee was later abolished but the gene-splicing group still exists.

Secondly, the Ministry of Education, on the initiative of the relevant Research Councils, started in 1984 a five-year programme for promoting research in molecular biological techniques. The scope of this programme was broader than the scope in the Danish Council of Technology programme described above. It included plant improvements and molecular biological methods in animal breeding. The programme has about 33 million Danish Krone at its disposal for the five year period.

Thirdly the Ministry of Agriculture in 1985 started its own five year research programme. The programme covers a broad range of activities - animal husbandry, plant breeding, production of food and new applications of biomass. The economic resources of the programme are about 27 million Danish Krone over five years.

Lastly in 1986, the Government introduced a proposal for a biotechnological research and development programme. The programme was accepted by the Folketing at the end of the parliamentary session in May 1987. Nearly all parties voted for the programme and its financial means amount to 500 million Danish Krone, or approximately US$ 70 million.

The proclaimed goals of the programme are to create a sort of "green revolution" in Denmark: to generate new industries based on the new "green technologies", to promote health care and to make Denmark the most advanced agricultural and fishing country of the world. Biotechnology is also expected to help solve the environmental problems from the "old" technologies used today. In practice, these goals are to be achieved through the establishment of biotechnology research centres, which in fact would appear to be cooperative agreements between universities and between firms and/or between research institutes and firms.

Who were the promoters of this programme? In November 1985, the Ministry of Education, the Ministry of Agriculture and the Ministry of Industry appointed a group consisting of three leading civil servants representing the interests of the three Ministries. Their proposal, dated January 1986, forms the basis of the governmental programme. I have as yet been unable to determine whether the promoters were the Ministers or their civil servants. It is known that the Minister of Education himself has been and still is involved in promoting new technology in general and biotechnology in particular.

The reason for the unusually long time in the parliamentary process of the governmental proposal of a programme of research and development in biotechnology is to be found in the fact that the majority in the Folketing has a much more explicitly formulated set of goals in relation to the programme than the government had. The Social Democratic Party put forward, shortly after the proposal from the three civil servants was made public, a proposal for a programme in biotechnological research and development. This proposal is largely identical to the governmental programme, but in one important respect it is different: its goals are more explicitly formulated in relation to health, nutrition and the lowering of prices for pharmaceuticals and food. One target is to the promotion of new ecological farms.

## TECHNOLOGY ASSESSMENT AND INFORMATION

In the governmental programme as well as the proposal from the Social Democratic Party, it was stated that it is important to undertake technology assessments and to provide information to the general public about biotechnology. About 4 percent of the total budget has been reserved to these two activities. The background for this interest in technology assessment and information about biotechnology can be found in the attitudes in the population. In March 1987, the Kasper Vilstrup Institute conducted an opinion poll in the adult population in Denmark in relation to gene-splicing. 22 percent said that, in their opinion gene-splicing was mainly useful for the society. 34 percent said that, in their opinion, gene-splicing was mainly harmful. 30 percent found that the technology would be useful as well as harmful. 14 percent had no opinion. The men were more positive in their answers than the women. Women said that the technology would be useful as well as harmful.

## THE DANISH TECHNOLOGY BOARD CONFERENCE ON GENETIC TECHNOLOGY : APRIL 1987

The Danish Technology Board and the Society of Danish Biologists held a consensus-conference in April 1987 on genetic engineering in industry and agriculture. The consensus-form has been used before in Denmark, but this conference was the first of its kind where laypeople themselves chose their experts, and limited the subject to what they found relevant to the issue. The idea was to give lay people as much influence as possible on the context and the outcome of the final draft from the conference.

The panel of inquiry was selected through advertising in local newspapers throughout the country. 14 laypeople with ages and from social backgrounds representative of the general population were chosen. During the planning period, the group was efficiently tutored, to make the panel of inquiry the leading force in preparing the inquiry agenda, defining the major questions and choosing and interrogating the experts.

In September 1986, a large circle of experts with different political and ethical opinions on genetic engineering and view on society were asked to put forward their views. In October 1986, the panel of inquiry was selected and the first preliminary seminar took place. The panel of inquiry was introduced to each other, and lectures on genetic engineering were given by a biologist. In January 1987, the second preliminary seminar took place. The panel was given further detailed information by 4 instructors, and they put forward their principal questions and selected their experts. Only then were the 14 selected experts contacted and requested to compose more detailed drafts on the subject. These drafts were studied by the panel and used for reference in the ensuing conference. On the first day of the conference, the experts lectured on their opinions and theories. After each presentation, the panel questioned the experts. During the following evening the panel summarized the questions that had not been answered and prepared the next day's inquiry.

On the 2nd day, the panel questioned the experts, after which the audience was given opportunity to comment on the issues. During the following evening and night their final draft and conclusions were composed. On the 3rd day, the conference received the final draft and debated it. A press conference concluded the week-end. A follow-up report on the whole process is presently written by 3 scientists from the Copenhagen School of Economics and Business Administration. The final draft and the lectures of the experts have been published by the Danish Technology Board.

The panel of inquiry composed their final draft, a 14-page document, during the night between the second and the third day of the conference. This is their own summary.

> Genetic engineering is different in quality from biotechnology formerly used in agriculture and industry because it has become possible to combine genes across the natural species barrier. Natural evolution is hereby eliminated.
>
> We believe, that this confronts our ethics with new challenges forcing us to attempt to formulate a view of Nature and to ask ourselves, what we - as human beings and as a part of the created world - can permit ourselves with regards to the natural world. We do not have the unconditional right to intervene, but we do implicitly have the duty to protect.
>
> A majority of the panel accepts genetic engineering on microorganisms, on condition that they are kept in closed compartments, under strict control and subject to precautionary measures, and that the use of genetic engineering is of service to mankind.
>
> A minority of the panel (2) believe that genetic engineering must be completely banned. They fear, that sooner or later, laws of limitation will not be respected, in which plants and higher animals could be implicated.
>
> About half of the panel believe that genetic engineering on plants must be prohibited, about one third will allow genetic engineering on plants, and about a sixth take no position.
>
> The panel agrees upon the fact that genetic engineering on animals must be prohibited.
>
> To our knowledge at present, research on risks in connection with genetic engineering is inadequate. We do not know, whether genetic engineering will cause disruptions in the eco-system. Risks assessments, that take all risks into account, are not possible. Our knowledge about ecosystems is too limited. Furthermore, it is difficult to agree on definitions of risk and danger.

As a result the following suggestions were made; that an ethical council be set up, consisting of laypeople only; that research on ecologically sound agriculture be intensified, emphasising the quality rather than quantity of production; that industry must be held responsible for any hazards created by their products; that more funds be made available for research on risk and on ecosystems; and that an independent information programme be created, enabling members of the public to understand genetic engineeing and to develop their views of it.

## SOME CONCLUSIONS AND SOME LESSONS FROM THE DANISH CASE

In the period 1976-1986, there has been a growing consciousness among both parliamentarians and civil servants in Denmark about the choices that can and must be made in the field of biotechnology. An indication of this growing consciousness was the passage of a special law by the majority in the Folketing to regulate the industrial application of gene-splicing. The interesting issue in this case was that nearly all the technical experts were against this special law, because they felt there was sufficient legal basis in existing law. Parliamentary concern set that aside. Rightly or wrongly, I consider it important, that parliamentarians themselves take decisions on issues that they or their voters regard as important and that inescapably involve essentially political judgements. The members of parliament also showed a clear understanding of the very different problems within the broad field of biotechnology. Important distinctions can be drawn between industrial fermentation in closed production systems, the release of gene-spliced microorganisms into the environment and genetic engineering in plants or agricultural animals. Finally, the ethical and political distinction between applied human genetics and other forms of biotechnology have been well understood.

An important reason for the parliamentary influence over developments in biotechnology as opposed to their influence over developments in other technologies can be found in a combination of two things. Firstly, the technical experts failed to understand the deep scepticism of most people in society, grounded in the scandals within the chemical industry and a concern with nuclear energy. During the last ten to fifteen years, there has been a growing scepticism about both the neutrality in the technical choices available to us and the integrity of the experts. Secondly, various groups in the population discovered that knowledge about the technology and its possibilities, combined with organizational work, could influence the decisions made in firms and in the public sector. Energy policy in Denmark since 1973 has been an analogous case, with success for non-nuclear development, even though there was a clear parliamentary majority in favour of a nuclear energy strategy prior to that. Today, this situation has changed at the Parliamentary level.

It is interesting to notice that the parliamentarians, in the case of biotechnology, made active efforts to obtain information from sources other than the established scientific system. Here, science journalists and technology assessment researchers have had an important influence in two respects. First, they have alternative information to the generally accepted view. Second, as a group, they have helped to put other things on the agenda, and shown that science policy could have other priorities.

Perhaps the most important feature of the new biotechnological research and development programme, regardless of whether it is the government version or the Social Democratic version, is the recognition of the necessity to make choices among all the possibilities opened up by this technology. If the state does not make decisions, the firms will. In Denmark the result would be the dominance of the pharmaceutical firms, since they possess both the technical expertise and the financial strength necessary to stay in business in this technology. It is questionable whether this strategy is good for overall Danish national interests with our traditional positions of strength in agriculture and food. But, and this is important, it has been accepted politically, that technology assessment and information for the population are necessary, if people are to accept the technology, and if the most obviously incorrect investments are to be avoided. This acceptance of some influence for the democratic political system, which is based on two different but actually convergent arguments, namely influence for the majority of the population as a democratic right and as a method of avoiding erroneous public investments, represents a very important shift in political life from the laissez-faire policy formerly accepted by nearly the entire political spectrum in Denmark. It is also interesting that this shift did not arise from within the bureaucracy, but within Parliament, especially from the left wing in the Folketing with some support from the centrist parties.

Lastly, it should be stressed, that the parliamentarians' decisions in relation to the biotechnology, putting into practice more restrictive legislative initiatives than thosee wanted by the industry and the scientific community are accurate reflection of popular opinion in Denmark these days. The result of these decisions will be, in the first instance, at first hand, a slower and more restrictive regulatory procedure in the field of biotechnology, which will be subject to criticism. But I think that in fact it will give Denmark a competitive position in the medium and in the long term, because the result will be to create institutions with strong competence in risk research in this field, and possibly a more highly informed and aware public. But I do not think that the result will be that people will change their ethical attitudes or come to judge differently the trade off between ethics and economic interests. I think that the use of microorganisms in recombinant DNA technology experiments will be unproblematic - with the restriction that such work is done with the necessary precautions in relation to possible risks to health and the environment. But in relation to the genetic manipulation of plants, animals and human beings, people will demand much more profound arguments than those based simply on scientific interest or commercial benefit.

# 12 Public Debate on Biotechnology: The Experience of the Bundestag Commission of Inquiry on the Opportunities and Risks of Genetic Engineering

Wolf-Michael Catenhusen

## INTRODUCTION

In June 1984 the Commission of Inquiry on the Opportunities and Risks of Genetic Technology, a special commission of the German Bundestag which sat until January 1987, became the first parliamentary committee in the world to try to define and evaluate the opportunities and risks associated with genetic engineering and to make recommendations to Parliament for political action in those areas where it found that there was a need for society to take action. The recommendations put forward range from proposals for promoting research to recommendations for legislative measures. The Commission of Inquiry's report was issued in January 1987, and in October of the same year the Bundestag started to discuss it in committee.

In my preliminary review of the Commission's work, I would like to do three things, from my standpoint as its chairman: to consider how the Bundestag came to take this unusual step; to present some important results of the Commission's work which will certainly be relevant to the debate in other Western European countries; to analyse the significance of this work for the increasing public debate on the opportunities and risks of genetic engineering in the Federal Republic of Germany.

## BACKGROUND TO THE ESTABLISHMENT OF THE COMMISSION

Until the beginning of the 1980s the establishment of genetic engineering research in the Federal Republic of Germany went ahead without attracting much public attention. The Federal Research Ministry had begun funding genetic engineering projects as early as 1974. In 1978, following the debate on safety aspects of genetic engineering in the United States, the Federal Research Ministry set up the Central Commission for Biological Safety (Zentrale Kommission für die biologische Sicherheit - ZKBS), a body composed of four scientists working in the field of genetic engineering, four scientists with special experience of safety aspects of biological work and four representatives from trade unions, industry, the industrial safety sector and research promotion organizations. The ZKBS's work is based on the "Guidelines on protection from dangers from nucleic acids recombined in vitro", adopted in 1978. Federal-funded research projects have to remain within these guidelines. The chemical industry trade association in the Federal Republic stated in 1981 that it would voluntarily observe the safety guidelines. A proposal to give the guidelines statutory force had the effect of widening the debate in 1978/79 by involving an interested section of the public for the first time, although Parliament did not take part. Massive opposition from many West German scientists, very soon caused the Social Democratic/Liberal coalition government to drop its legislative plans.

The first major investment by a German chemical group, the Hoechst company, in the USA - the establishment of a genetic engineering research centre in Boston in 1981 - rapidly led to massive support from the state and industry for genetic engineering, particularly for projects and undertakings likely to bring about a closer link between basic research and industrial application. One example was the creation of four national genetic centres, with participation by the major chemical companies. This action demonstrated that the politicians now realized that genetic engineering was a "key technology"; it had received the accolade of being regarded as "necessary" for the future industrial prosperity of the Federal Republic and for its international competitiveness. The stepping up of assistance from the state intensified the public debate about the application of the new technology, the opportunities it presented and the risks involved. The debate was also a reaction to the great expectations associated with genetic engineering. Initially, discussions were limited to a small circle, mainly within academic institutions linked to the churches, although the trade unions were also involved. The prospect of the future application of genetic engineering to human beings (genome analysis and genetic therapy) very soon came to the forefront of the debate, stimulated by events in the USA, where Congress was then holding important hearings on the implications of genetic engineering.

Increased parliamentary activity in Germany, leading to the appointment of the Commission of Inquiry on the Opportunities and Risks of Genetic Engineering, thus could not be said to have been merely a response to a wide and heated public debate that was already under way. In 1984 Parliament had not yet been forced to act, environmentalists and the women's movement had not yet taken the subject up, and the political parties had yet to adopt a position on it. Thus it was possible for a small group of MPs from all parties, except the Green Party, rapidly to reach an informal consensus on the appointment of the Commission and the assignment of tasks to it. In my view the motives behind this move were as follows. Firstly the few MPs who were interested in the subject were deeply concerned by the breathtaking speed at which genetic engineering research was developing. To avoid a repetition of the mistake s made in the nuclear energy debate, they all thought it would be useful to evaluate the new technology, which now stood at the threshold of commercial exploitation, and to try to assess its future impact. Secondly there was agreement that a commission of this kind could be a suitable source of information for Parliament. Thirdly there was broad agreement among the MPs involved that conceivable future applications of the technology to human beings and increasing commercial exploitation for various purposes raised the question of where the boundaries to the application of the technology should be set. It was felt that this was a question which should not be left to science and industry alone but should be the responsibility of the state.

In setting up the Commission, one of the prime concerns was to attract competent representatives from different positions in industry, science, the trade unions and the churches, since its members were to be both MPs and non-parliamentary experts, all with equal voting rights. The concrete tasks assigned to the Commission quickly silenced objectors in the conservative camp who claimed that technology assessment would be used by the Commission as a means of "technology arrestment".

The Commission's brief was "to assess the opportunities and risks of genetic engineering in the principal applications currently emerging, especially in the spheres of health care, food, raw material, energy and environmental protection, with particular emphasis on economic, ecological, legal and social repercussions and safety aspects". In addition, the Commission was required to propose concrete measures to be taken by Parliament.

## IMPORTANT RESULTS OF THE WORK OF THE COMMISSION

In January 1987 the Commission issued a report which had been approved by all the experts except the representative of the Green Party.

The Commission agreed that our expectations in connection with the development of genetic engineering must be realistic, and that the mistaken changes of course made in connection with other technologies when they were still in the "euphoria phase" must be avoided. For example, genetic engineering gives no prospect of any improvement in the employment situation in our country in the foreseeable future which would make promotion of the technology an urgent priority. Even in the medium term, biotechnology and genetic engineering will not replace oil-based chemical synthesis as the most important method of making organic products. In the pharmaceutical industry genetic engineering may make it possible to optimize existing biotechnological processes, thus giving rise to an innovatory advance within existing product ranges. But only in a relatively narrow sector of activity will there be any start-ups of highly innovative small firms specializing in developing, and in some cases also producing, high-value diagnostic and pharmaceutical products.

It is already apparent from the application of genetic engineering in many fields that it will not be the dominant or exclusive technology for producing individual services for society, either in medicine, the chemical-pharmaceutical industry, crop-growing, livestock farming or the production of energy from biological sources. In many fields genetic engineering, despite being a new technical option, is having the effect of maintaining existing social trends, although it may be strengthening or weakening them. In this connection public criticism of possible applications of genetic engineering is directed simultaneously or primarily against over-arching strategies which have developed independently of genetic engineering. What people are really worried about perhaps is the basic problem of the industrialization of agriculture with its attendant pollution of the environment with increasing effects on human lifestyles. Greater use of genetic engineering will in no way relieve us of the necessity to pursue more vigorous policies in defence of the environment or to take action to reduce the intensiveness of agriculture with its related environmental problems. The Commission also agrees, for example, that the major problems of the Third World, including nutrition and health, are primarily political, social and economic in nature and must be tackled on that level. Nevertheless, genetic engineering can make some very useful contributions to solving social problems in many fields, such as early diagnosis and treatment of major diseases, combating hunger in the Third World, improving the quality of our food and developing more environmentally acceptable production processes.

The debate about safety in genetic engineering research is as old as genetic engineering itself and has led, following the Asilomar Conference of 1975, to the introduction of safety guidelines intended to ensure the safety of work with biological organisms through a system of biological and physical safety measures. The Commission takes the view that the system - a combination of physical and biological containment - has proved its basic worth. All experience to date contradicts the original assumption that the recombination of nucleic acids was liable to give rise to completely new, hitherto unforeseeable and indefinable dangers. This experience can be cautiously extrapolated, at least to strains and processes similar to those already used in research. Neverthless, the transposition of experience gained with biological agents at the laboratory scale to the scale required for production ("scaling-up") poses new safety problems. These can and must be dealt with through further development of the existing safety philosophy.

The principle of biological and physical containment obviously cannot be applied to the selective release of genetically engineered living organisms into the environment. In this case the survival of the released organisms is the very condition required for their successful use. The precise effects of such releases on ecological balances are not yet known, and not only because virtually no attempt to do so has yet been made. The Commission has made a detailed study of the problems of biological risks, in particular the question of the ability of organisms to survive and reproduce, their ability to exchange genes with other organisms and their dispersal capability. In this sphere, the risks inherent in field trials involving cultivated plants and domestic animals, wild plants and wild animals, viruses and micro-organisms, vary very widely. On the other hand, we still have virtually no information or test procedures which would enable us to make concrete observations and assessments of all irreversible release experiments ( for instance on reproduction and dispersal). It is precisely this lack of knowledge which has prompted the Commission to consider temporary bans in this area also. They would apply, for instance, to micro-organisms into which new hereditary information had been implanted by means of genetic engineering. We wish to find out whether safety research and ecological research can lead to a more accurate assessment of the consequences for human beings and the environment. In this connection, the question of what residual risks we in the Federal Republic are willing to accept in irreversible interventions in the ecosystem will have to be clarified at some future time. Such a decision would, however, have to involve the Bundestag.

In many cases the use of genetic engineering can bring benefits; but without an appropriate regulatory framework it could also maintain or strengthen undesirable social trends. Genetic analysis can be helpful in the prevention or early detection of occupational diseases, but it could also result in the non-appointment of a job applicant or the loss of a job. For this reason the Commission, on principle, rejects the general application of genetic analysis to mass screening of workers. It supports the use of genetic analysis of workers only as part of an industrial health care programme, and then only if the recognizable dangers of its misuse and the likelihood of adverse developments in the industrial safety and social insurance systems can be reliably prevented or avoided by legally binding regulations. This must be achieved by measures including statutory restriction on the scope of an employer's questions at job interviews and by a ban on electronic storage of genetic data.

In the near future genetic engineering will come progressively into practical use. Therefore a compulsory framework must be created to regulate relevant research, development and applications. The Commission proposes that generally binding safety regulations for genetic research establishments and corresponding production units should be statutorily imposed in order to protect human beings, animals and the environment. The regulations should however be flexibly adapted to the current state of the art. Linked to this we propose the introduction of hazard liability for genetic engineering projects requiring a permit, because in some areas of application the level of scientific and technical attainment is not yet backed by sufficient experience, and the relevant regulations are necessarily provisional in character, thus leaving a possible residual risk. In addition, deliberate infringmement of the safety regulations should be made a punishable offence.

The Commission proposes that restrictions be imposed on genetic engineering in three important areas.

The Commission proposes unanimously that "germ line gene therapy" with human genes, which may become possible in the future, should be made a criminal offence, because, among other things, it might open the way for selective breeding of human beings. On the other hand, the Commission considers interference with cells in the human body (somatic cell gene therapy) to be a basically acceptable form of therapy.

The Commission proposes that genetic engineering research projects should not be carried out in military establishments in our country or financed from the defence budget. An exception would be made for research work which by its nature comes under the heading of military medicine. The public should be kept fully informed about any projects of this kind.

The Commission proposes five-year bans on some of the environmental release experiments being contemplated.

The report calls for the use of the whole available range of practicable instruments to ensure that genetic engineering in a responsible way. We would like this framework to be established before genetic engineering comes into widespread use in our society, thus indicating how the opportunities can be used responsibly and in a manner acceptable to society, how proper care can be exercised in the face of possible risks and how recognisable dangers can be excluded to the greatest possible extent. This could help to prevent any polarization of society into pro- and anti- genetic engineering camps. This consensus, which was supported by all the expert members and by the representatives of the SPD, CDU/CSU and FDP, was unexpected in view of the experience with other technology debates. The timing of the appointment of the Commission - before extensive public controversy had arisen - may have been a contributory factor.

Our report will stimulate further public debate, not bring it to an end. Genetic engineering will continue to be a subject for technology assessment. For this reason the Commission, mindful of its objectives and to save unnecessary work, has decide not to try to classify the technical potential of genetic engineering in terms of different models and development scenarios in our country's future industrial and economic structure. Moreover, too little concrete information is available, even from the development of genetic engineering to date, to enable us to determine whether - and if so, how - genetic engineering will decisively affect such structures. However, these are problems which, as the technology develops and is applied, should be considered by the Bundestag as part of its technology assessment projects.

The debate about the opportunities and risks of genetic engineering, which is intensifying, will always be bound up with the debate about the way we want to live in the future. The questions of principle associated with this will have to be settled, but this could not be achieved by a commission of inquiry on "the opportunities and risks of genetic engineering". We shall be discussing them in the churches as well as in Parliament.

## THE REPORT OF THE COMMISSION AND THE PUBLIC DEBATE ON GENETIC ENGINEERING IN THE FEDERAL REPUBLIC OF GERMANY

In parallel with the work of the Commission of Inquiry the public debate on the opportunities and risks of genetic engineering has widened considerably, in all political parties, the churches, the environmental movement and the women's movement. It often overlaps with questions of reproductive biology and in vitro fertilization. The Commission's work and findings have greatly influenced the positions taken up by the political parties and other organizations in the genetic debate. I do not exclude the possibility that this influence is evident even within the Green Party, where regional associations and individual MPs have been modifying their position of unequivocal opposition to genetic engineering. An initial campaign has brought together a coalition of environmental associations extending as far as church groups, to tackle the problem of the use of genetically engineered bovine growth hormone.

The lines along which the Commisstion has worked have been dictated by the conditions governing parliamentary work, with the Commission concentrating on preparing decisions for Parliament and thus being heavily involved with regulatory matters. Here the Commission has been able, firstly, to make successful soundings to discover the areas where proposals for political action could be formulated, so as to win a broad consensus of approval. Secondly, the Commission's work has had the merit of putting Parliament into a position in which it can now take decisions in the proper context on a large number of regulatory matters arising in connection with the progress made by basic research in genetic engineering and the transition to commercial exploitation. This applies from patent law, through the need for our own genetic engineering statute, right through to drafting the approvals procedure for the marketing of genetically engineered medicines and the problems associated with the release of organisms into the environment.

Even for many critics of genetic engineering, the section of the comprehensive report dealing with the current state of the art provides an important foundation for further public debate, because it is the first work to deal with all the important assessment problems of genetic engineering, including the question of possible military applications.

Right at the beginning of the Commission's work there was agreement among its members - apart from the Green Party representative - that it was not possible to make any blanket assessment of genetic engineering. Questions of the ecological, social or ethical and moral acceptability of genetic engineering diverge very widely, if one thinks of laboratory research, the production of genetically engineered medicines with the aid of bacteria, the release of genetically modified micro-organisms, the breeding of herbicide-resistant plants or the selective breeding of human beings with the aid of "germ line therapy". This consensus, which tended to grow even stronger as the Commission's work progressed, is naturally open to criticism from those who see genetic engineering as a step by mankind beyond its permitted limits. For these people genetic engineering, like the splitting of the atom, proves that modern science has taken a wrong turning and is heading in a direction which threatens the human race. To preserve human dignity and safeguard natuaral evolution this development must be stopped, it is felt, even at the cost of prohibiting certain scientific work. However, the great majority of the Commission's members thought that these questions were ouside the scope of its work. A short-term parliamentary commission cannot set itself to act as an arbiter of cultural and social change. These sorts of criticism of genetic engineering are often simultaneously or primarily criticisms of over-arching strategies which have developed independently of genetic engineering, strategies involving basic problems of the industrialization of agriculture, pollution of the environment by our industrial system, etc.

In October 1987 the Bundestag started to discuss the report in twelve committees. So far the CDU/CSU, SPD and FDP have been using the Commission of Inquiry's report as the basis for their work, although a number of the report's proposals have come in for criticism from scientific organizations. Their main objections are to the idea of making the safety guidelines mandatory and to the proposed restrictions on experiments involving the release of genetically engineered organisms into the environment. So far the Federal Government has not commented comprehensively on the report. A number of parliamentarians fear that national regulations limiting the application of genetic engineering or subjecting it to stringent safety standards could very soon be circumvented by European Community regulations drafted on the principle of the "lowest common denominator".

The Commission's work was an experiment. The Commission proposes that genetic engineering, both research and application, be placed within a legislative framework before many of the possible technical applications already foreseeable become a reality. This could be the right way to make use of the opportunities offered by genetic engineering in a responsible way which is acceptable to society, to take proper precautions against possible risks and to say "no" to ethically unacceptable applications. It could also help to prevent society becoming polarized into pro- and anti-gentic engineering camps as happened with nuclear energy as a result of political failure.

# PART SIX: CONCLUSIONS

# 13 Learning about Participation in Biotechnology

Edward Yoxen, Vittorio Di Martino

## NOTIONS OF PROGRESS

The preceding twelve papers display a range of different attitudes to biotechnology. All are serious commentaries on some aspect of the field. All are cast in the form required for the meeting. They are short, technically accessible, and indicate where more research is needed. Some are more theoretical than others: some more closely tied to government policy. They follow a standard pattern. Yet each is distinctive in some way, displaying not merely the particular expertise or specialisation of the author, but also giving some indication of his or her underlying commitments, preferences and assumptions. Two papers at least, those by Mark Cantley and Peter Daly, show real and uncomplicated enthusiasm for the technology, in the belief that innovation will translate straightforwardly into tangible benefit. On the other hand in two papers, those by Nadine Fresco and Gerd Junne, there are signs of profound scepticism about the value of particular aspects of biotechnology, given the economic, social and psychological problems which will ensue. Their comments are by no means outright condemnation, but a careful statement that serious questions will arise and should be anticipated now.

The remaining papers are written in such a way as to preclude any such decoding, although one might reasonably infer that all the authors regard biotechnology as a technical and industrial phenomenon of some importance, the realisation of which will require planning, analysis informed by economic and sociological theory to some degree, evaluation and political negotiation. The very fact that each author has invested time and effort interpreting the available data, or describing the relevant economic models or analysing particular political initiatives suggest that they believe such reflection is generally useful. They recognise that technologies are not simply created by some magic of the market; they can only emerge through tortuous processes of invention, promotion, appraisal, negotiation, and endorsement.

Not everyone shares this technocratic optimism that, given enough data, enough anticipatory modelling, and enough discussion, a tolerable, even an exciting, future can be engineered through biotechnology. Some people would simply like industrialised societies to forego biotechnology, either because it is held to be too risky or because it is thought to extend an intolerable system of economic domination. One problem with that view is where you draw the line. Which biotechnologies are unacceptable, to whom and why ? Another is practicality. How is the societal self-denial of whole areas of technology to be arranged ? Accordingly it seems much more realistic politically to develop the institutions which would allow more debate and selection to take place over the politics of technological innovation.

Attitudes have shifted a great deal over the last fifteen years. In the mid 1970s, when the invention of procedures to create recombinant DNA molecules first excited such scientific and industrial interest, there was real concern with the possible medical hazards such molecules might pose in the workplace. So seriously was this hypothetical prospect taken, that a whole category of experiments was deferred for a limited period. After some reflection many scientists adopted the view that such caution was unnecessary, and the field of recombinant DNA research expanded with extraordinary speed. One result was the burgeoning of industrial interest in the convergent and interacting set of disciplines, that we now call biotechnology. Another was the gradual diffusion of the view that any risks posed by genetically engineered organisms in the workplace could easily be managed, by insisting on good laboratory practice or intelligent safety engineering in the design of new industrial plant. Despite the existence of certain anomalies, such as unexplained outbreaks of disease in particular labs, which could derive from unrecognised hazards when using viruses and bacteria, and despite evidence that scientists often fail privately to observe agreed safety guidelines, the orthodox view today is that the maintenance of safety in biotechnology facilities is not a serious issue, and need no longer be politically controversial. In the view of many, as a technical and political problem it has been solved.

But if the few commentators who remain sceptical that such definitive conclusions can yet be drawn about lab safety would find it hard nowadays to attract much attention in the media or in scientific fora, the same is not true of those who speak of risks to the environment. It is somewhat ironic, or depressing or intriguing, depending on one's point of view, that an issue that was set on one side in the 1970s as of secondary importance should now have such political salience. For it is clear that the prospect of the ever more frequent release of genetically engineered organisms - viruses, bacteria, plants, and insects - into the environment has evoked a new wave of concern about possible risks. The fact that such releases are inevitably experimental at this time and that their consequences could be both serious and irreversible is being interpreted by a range of environmentalist organisations as a matter of real concern, which ought to lead to another moratorium.

But this is just one form of environmentalist opposition to initiatives that are central to the programme of biotechnology. Organisations and political parties which prioritise environmental issues are also voicing concern about the extent to which biotechnological innovation entails or is ineluctably drawn into the intensification of animal and plant agriculture. At a time when European agriculture is being managed in a way that encourages over-production and when the pressures to increasing industrialisation of farming make the position of small producers ever more precarious, these critical sentiments command significant attention. Although biotechnology could be used to promote resource-conserving, ecologically rational, low energy agriculture, many suppose or assume that it will in fact intensify the existing deleterious trends. Their real concern about the economic impact of existing agricultural policies translates into antagonism to biotechnology.

This note was sounded repeatedly at the Dublin seminar by Green members of the European Parliament, using a political idiom that was at the same time both conservative, evoking an expert community of traditional producers, sensitive to nature, and threatened by urbanisation, science, and a modernising state bureaucracy, and progressive, being anti-capitalist, oriented to real human needs and opposed to the alienation of modern productive activity. Not surprisingly this anti-modernism draws its political support from sections of the professional middle class, radicalised by student politics but distanced in some way from the labour movement and from small producers in rural communities, who feel betrayed by the traditional conservative parties.

At the same time the medical applications of biotechnology also evince mixed reactions. Whilst the discussion of such possibilities now generally acknowledges that all kinds of technical, financial and moral difficulties may well arise in commercialising new therapeutic and diagnostic products, and that their impact on health may be less striking when considered in a national context, nonetheless medical innovation is still widely regarded as intrinsically worthwhile.

Although the pricing, marketing and sales policies of pharmaceutical companies are often the subject of strong criticism, and significant sections of the European population use traditional or alternative medicine, the goals of producing new medicines, drugs and medical devices through the continuation of medical research is generally seen as legitimate, despite the clear evidence of concern about and opposition to aspects of medical practice. Very often what doctors do, which can excite criticism, such as taking decisions for their patients, is effectively decoupled from biomedical research, even though it would enable them to behave in the same way in new areas.

Only in the fields of reproduction and human genetics, the latter a rapidly expanding area of clinical science, does this conventional compartmentalisation of the problematic and the unproblematic break down, possibly because the medical procedures are largely experimental at this time. Also the issues involved touch upon very basic questions of human identity and value, such as the nature and sanctity of human life, the moral claims that embryos and fetuses may make upon adult human beings, the nature and significance of physical and mental handicap and the connection between morality and the law. Here as one might suppose real differences are emerging over what counts as progress. Procedures such as the antenatal diagnosis of genetic diseases and chromosomal disorders are encountering more critical comment, and more endorsement, as they become more widely used and better known. The criticism takes several different forms including that from some feminists, who see a concern with the prevention of genetic disease as an oppressive ideology conveying distorted or bogus eugenic ideas of human perfection and as entailing a further erosion of women's power over their health and reproduction. A significantly different kind of criticism of antenatal diagnosis has been voiced by various Churches as a negation of the sanctity of human life.

These remarks simplify drastically a whole range of complex attitudes, some of which overlap, and some of which are counterposed, but it is interesting to note that, as with the environment, criticisms of what is conventionally counted as progress are emerging both from groups that one could characterise as politically progressive as from those that tend to be politically conservative. Conventionally also such issues are often bracketed off from the mainstream of industrial biotechnology, as a special set of moral problems, that should be considered on their own. At one level this cannot be denied; but at another it is highly misleading, since there are important technical and commercial connections between industrial microbiology and the new human genetics. Moreover it would be idle to pretend that significant moral questions are not raised by other aspects of biotechnology and its applications, for example the conservation or destruction of rare plant species or the long-term environmental consequences of changes in land use.

## INFORMATION AND NEGOTIATION

The Dublin seminar was attended by some seventy people, all of whom had some interest in or professional contact with biotechnology. Although the intention was primarily to develop and discuss future research directions, the academic participants formed only a small minority. Representatives of trades unions, of industrial management, of international organisations, including the European Commission, officials of national government departments and politicians outnumbered them massively. One might reasonably suppose that such participants hold rather divergent views on a number of policy issues, and that some, expecting such differences of view, came expecting disagreement and positional debate. In the event, although some conflicts of opinion did emerge, it was perhaps more striking that many people felt they had been made more aware of the underlying issues and would be more attentive to biotechnology in future. This emerged for example from the comments made on forms returned to the Foundation after the meeting. It is not that people necessarily changed their minds, but that some realised the complexity and difficulty of the issues at stake, and that more investigation and information is essential, if the conflicts of interest are to be moderated and used productively, if not actually removed.

It was also apparent that many people accepted that innovation in this area would require some degree of negotiation, for example to set standards, to set prices, to obtain development support, to secure access to raw materials or to establish a new market. One might say that the attendance of people from a wide range of organisations was some testimony both to the fact that biotechnology is viewed as a technology of growing social import, and to the fact positions on the issues it raises must be researched, debated, and refined. People came to gather information, to test arguments and to survey the changing political landscape. Manifestly this was the case with some of the more influential participants, who contributed frequently to the public discussions, but it must also have been true of those who said less.

This state of affairs defines both a challenge and an opportunity for the European Foundation for the Improvement of Living and Working Conditions. It is a challenge in that prospective, policy-oriented research is not easy to formulate, prioritise or carry out, particularly in a fast-moving, highly interconnected area of technology. Moreover there may not be easy agreement over what counts as useful and timely research. But a considerable number of suggestions for research were collected at the seminar. There is then a remarkable opportunity to take up some of them and to develop an area, which in Europe at least, remains largely unexplored.

The review papers above provided the basis for discussion in four working groups on the first day of the seminar. On the second day the ideas for research that had emerged were presented to a plenary session for review. These are described below, under the headings defined by the workshop subjects. Certain general themes also are also apparent and these too are discussed below. In particular it is clear that enough has now been written in a speculative mode, to allow and to require the generation of scenarios and hypotheses which could actually be tested against social reality.

## RESEARCH ON BIOTECHNOLOGY AND HEALTH CARE

There was little doubt expressed by anyone in this working group that biotechnology would have a major technical impact on medicine over the next decade. To some extent this is already clear, with the appearance of new therapeutic products, new kinds of vaccine, new diagnostic tests and equipment, made possible by developments in molecular genetics, immunology and biochemistry. Moreover, as Peter Daly demonstrates in his review, several biotechnologies have a role to play in the development of conventional processes of pharmaceutical manufacture and in the design of new products. At the same time medical innovations of all kinds face a much greater degree of economic and regulatory scrutiny than was once the case, and for this and other reasons, the number of new products emerging from the new biotechnology companies has been less than expected ten years ago. Also as both David Banta and Nadine Fresco make clear in both public health and medical genetics there are major questions still to be resolved about the desirability of certain kinds of intervention in people's lives at various points. Simply because the technology exists does not mean that it is acceptable to use it, or that people want it to be used, or have even been given the opportunity to reflect on its implications and consider whether it should be used. Accordingly five areas were proposed for more research, relating in different ways to biomedical innovation. The first concerns attitudes to regulation, to ensure medical product and device safety. What determines attitudes to risk and its acceptability ? It is often said that people have much higher expectations of medical regulation than they seem to expect in other areas of life. Whilst this tends to ignore the distinction between self-imposed and involuntary risks, and to beg the question of what is an acceptable risk, if one enters a risky situation of one's own free will, there is an important issue here of how frameworks of understanding, that make possible judgements about risks, are constructed. How do we develop certain kinds of expectation, that medicines should be 'safe', that certain environments, such as hospitals, are likely to be safer (or not), that expert practitioners are indeed competent, as they claim to be, and so on. Here the cultural variation within the European Community is a major asset in research terms at least, for it allows all sorts of cross-cultural comparisons of attitudes to risk to be considered. But it is important to state that the mere compilation of data about different attitudes is likely to be an expensive waste of time. It is far more interesting and valuable to examine how such attitudes, beliefs and assumptions have emerged, are modified, interact and are replaced, within particular cultures.

The second area concerns the shifting relations between private investment and public health. For example many of the more highly publicised ventures in biotechnology have involved private capital invested in the commercialisation of new health care products, to be consumed either by publicly financed hospital systems or by privately financed hospitals that sell their services to insurance companies. The particular mix of public provision and individual consumption is often a result of political and economic struggles carried on over many years, and still very much an issue. Simplifying greatly, one can say that in Europe the supply of goods to the health care system is largely carried on by private concerns, whereas the supply of health care services to individuals is financed from public funds, to a significant extent, sometimes almost completely, and the regulation of health care costs and standards by the State is taken as a political necessity.

But it is very much an open question as to how the structure of investment in new technologies relates to health at the national level. Which systems are effective in other words in making a real and important difference to general mortality and morbidity, rather than simply generating a display of new products, of minimal value judged against the total burden of disease ? In addition to this one can also say that certain areas of medical technology are likely to be underfunded, as they have been in the past, because the anticipated rate of return is insufficient to attract investors. An example is vaccines, which used to be expensive to develop and could only be sold at a very low unit price. As a result only philanthropic or subsidised concerns entered this market. Whether this will remain the case, with newer, much more expensive vaccines is unclear. The general issue here is to consider whether the kinds of products that will inevitably be prioritised in times of major restructuring of health care systems will have a major impact on health and the quality of life, and if so, through what means.

This leads to the third area identified in this working group, and highlighted by Daly in his background paper, namely the effects of cost containment on medical innovation. Historically the scientification of medicine has allowed doctors within hospitals to increase their professional power and status by managing a slowly expanding clinical team and by utilising a range of ancillary technical services. One result has been an increasingly expensive hospital system, and a sharp division between hospitals and primary, community-based care. This economic system is now under enormous strain in many industrialised countries. Amongst other things this is likely to bring major reorganisation of technical services and in working conditions in such areas. It is also likely to see increased pressure on individuals to fund and manage aspects of their health care, including the performance of simple diagnoses. An example is the growth of self-diagnostic kits, made possible by biotechnology, for a range of conditions. Whether this will reduce the burden on hospital laboratories and general practitioners is hard to predict. It may have the opposite effect. Either way it is an example of the kind of cost-limiting innovation in the health field, the impact of which needs careful evaluation.

Fourthly there are many questions raised by the emergence of so-called predictive medicine, in particular the increasing knowledge of susceptibility to particular conditions, such as heart disease, diabetes and certain forms of cancer. Whilst it seems useful in principle to have foreknowledge of such possibilities, this may well not be the case in practice, if the information is poorly presented, or used to stigmatise and discriminate against the individual concerned, or is grossly misinterpreted by others. Information of this kind is frequently thought relevant to prospective or continuing employment. It could be sensibly used in such a context or it could be abused. More work needs to be done, to anticipate future possibilities here, to review the kinds of legal protection available and to consider what new controls are necessary.

Finally there is the question of eugenics. Competing definitions of eugenics were debated in the workshop. Whereas some people applied the term to the now distant extreme practices of the 1920s and 1930s, involving the sterilisation of those considered genetically inferior, and claimed that eugenics was now simply of historic interest, others argued that it always existed in many forms, and that some endure today and, rightly or wrongly, are generally regarded as unproblematic. Thus the antenatal diagnosis of genetic disease is often presented as being simply the generation of information of use to a couple expecting a child. But its effects, particularly when organised as a part of a general programme of antenatal care go beyond the individuals centrally involved, with a longer-term impact on social attitudes and practices. Also the allocation of resources to such programmes may well be justified in essentially eugenic terms, of the economic costs saved by preventing the existence of affected individuals. It is simply myopic to claim that eugenics is a thing of the past. There are many important issues raised by the rapid expansion of human genetics within medicine, for the evaluation of which much more sociological data is needed. Here again the cultural and historical diversity of Europe is an advantage.

## RESEARCH ON BIOTECHNOLOGY AND AGRICULTURE.

The agriculture working group was outstandingly productive of suggestions for future research. They asked that their proposals be prefaced with two general remarks, that biotechnological innovation does not take place in a scientific desert, and that its agricultural forms result from the interaction of the technological and price systems. The significance of the former is that developments in other fields, solid state physics, computing, metallurgy, theoretical chemistry, for example, may possibly facilitate new developments in biotechnology, thus complicating the predictive exercise. The import of the latter is that the incentive to develop new options is highly dependent on conditions in particular agricultural commodity markets.

The first area of interest is that concerning differential productivity and the factors favouring economies of scope and scale. This is of course a classical economic question, but it is highly relevant to biotechnology in agriculture. For example it would be easy to create contrasting scenarios for Europe, where on the one hand the general trend towards very large-scale, highly mechanised farming is continued, and on the other new opportunities for small-scale, less intensive, diversified, resource-conserving production appear. Thus it is important to ask how different kinds of agricultural production have become economically dominant, in order to consider what options are likely to be open in the future, as new biotechnologies become available. Which kinds of farm, farmer, and crop stand to gain most from changes in technology and in productivity ?

The second area follows from the first and concerns the extent to which farmers have become dependent on suppliers of farm equipment and chemicals and on large customers. Are the ways in which people use their land largely determined by the supply of essential inputs and by the needs of those that they supply in turn ? Clearly in some areas this kind of dependence exists, particularly where farms have been turned over to the highly intensive production of a single crop, such as wheat, or where the only customers are processors with the power to define the terms on which they will purchase the crop, such as peas for freezing. It is often suggested that biotechnology will increase this dependence, although this is clearly not necessarily true, and needs to be examined carefully in different European agricultural contexts.

This in turn raises questions about the kinds of economic power that processors of and traders in agricultural products can exert. These concerns are not themselves fixed entities. They face constant competition and must evolve as organisations and as users of technology. As they change they are likely to acquire greater economic power as purchasers, if, for example, their ability to use a range of different raw materials is enhanced, such that they can buy at the lowest price. One piece of exploratory research for the European Foundation by Gerd Junne and his colleagues in Amsterdam suggested that as the dairy industry becomes more concentrated and more specialised its influence over and ability to select amongst producers will increase. They also argued that similar trends are evident in starch processing, where one can predict major problems in the future for potato growers in the Netherlands.

Fourthly the whole area of employment, working conditions and training in farming and related industries is obviously very important, precisely because the nature, purpose and scope of agricultural production is so evidently in flux. For many farmers in Europe the question is now whether they can diversify into new areas. This is partly a question of finance, partly one of their resources of land and equipment, and partly one of confidence, flexibility and skill. These constraints operate differently in different areas, with different levels of education and different regimes of government assistance. Since biotechnology could in principle offer many new opportunities for diversification and change of practice, it is important to consider whether people are in a position to take up such opportunities or whether they are only of value to a minority.

Finally there are the questions of genetic diversity and interaction with the environment. Manifestly some forms of intensive agriculture and marketing can easily accelerate the loss of traditional varieties and can place the survival of ecosystems in jeopardy. Here also diagnoses and prognoses differ markedly, with some commentators arguing that problems such as the reduction in the number of species under cultivation are exaggerated or are at least manageable with adequate gene conservation programmes, and others arguing that far too little attention is being given to them, with the prospect of major catastrophes in the longer term. These matters do not relate directly to living and working conditions, although there are connections of an indirect kind. Thus in a real sense the process of structural change in the nature and purpose of agriculture in the developed countries concerns the kinds of crops grown, the kinds of materials used to protect them and stimulate their growth, the sort of machinery used to harvest them and the ways in which they are processed and their residues disposed of. These issues directly affect the appearance of the countryside, its accessibility for walking and sightseeing, the risks from chemicals used in agriculture, and the kind of work that is available to people in rural communities. The transformation of our interaction with living nature in agricultural production is also a transformation of the range of possible social experience; as we re-make nature we also re-constitute ourselves.

## RESEARCH ON BIOTECHNOLOGY AND WORK AND EMPLOYMENT

When high-technology innovation gives rise to new companies, even a new industrial sector, towards the end of a period of world-wide recession, two questions inevitably come to the fore. Will the work-places be safe, or will unknown hazards emerge ? And how many jobs will be created ? Thus the assembly of highly integrated microelectronic circuitry has turned out to be not insignificantly risky for workers, because the use of highly toxic and biologically potent chemicals in chip manufacture. New technology does mean new risks here. And whilst microelectronics manufacture and application has generated new employment, the net effect on the numbers in the workforce has probably not been great, given the numbers of jobs destroyed in the late 1970s and early 1980s.

As the exploratory research done for the Foundation in 1986 showed, these questions are not easy to answer, when asked of biotechnology. The data is simply not available, although intelligent speculation is certainly possible. The workshop on this topic produced the suggestion that a survey of working conditions in the 'new biotechnology firms' would be worthwhile, since we cannot simply assume that small, financially vulnerable, research-driven companies will necessarily have the resources or wish to devote them to safety engineering and the maintenance of safe working practices. This assumes that the risks are largely known already, which is not in fact the case. It was also suggested that measures be taken to secure better co-ordination of understanding, to ensure that emergent expertise in one company or laboratory or government department is easily available to other parties, and that guidelines and procedures are frequently compared and discussed. This is not just a question of monitorring what people do at the lab bench or in a chemical plant, but also a matter of ensuring that the highest practicable degree of protection is designed into the production technology, as it develops.

Regulation is only possible if there are enough, adequately trained inspectors, to monitor working conditions. The novelty of much of biotechnology raises the question of what skills are necessary for inspectors to permit them to work effectively. Some of them will require training to this end. It is important to ask how is this being done in practice and whether the results are acceptable. Similarly workers in biotechnology-based companies will need training in the safety aspects of new technology, as will their trades union representatives who will be called on to articulate and protect their interests, as also will managers, who have themselves to learn of their particular responsibilities. It would be quite wrong just to assume that work hazards of biotechnological processes will take care of themselves: only through careful research, training, negotiation and recognition of the political lessons of past difficulties can the necessary standards for a new kind of production be developed.

The other area considered in this workshop was the potential for job creation, job displacement and job destruction. It is clear at least that biotechnology will not lead to an immediate and dramatic expansion of demand for labour. Any effects are likely to be marginal, except in markets for highly specialised skills, developed through extended university training and/or years of practical experience. For the foreseeable future the numbers of people involved in advanced, research-based industrial biotechnology will be small relative to the numbers in manufacturing industry as a whole. What is not so clear is whether one of the effects of biotechnological innovation in, say, food processing or the chemical industry will be the destruction of existing jobs, as processes are abandoned as uneconomic or companies go out of business.

With this difficulty of prediction in mind, it seems worthwhile instead to focus on organisational changes in companies, which are both significant in themselves and which are likely to influence how any work is organised in future. There is a growing literature from industrial economists, sociologists of work and management theorists about a trend to the so-called flexible firm, with a workforce divided into peripheral and core groups. This is referred to in the paper by Edward Yoxen above. As an idea it is both empirically under-researched and theoretically ambiguous, but it is sufficiently plausible and important for one to consider whether there is evidence of a trend in this direction in major companies involved in biotechnology, in food processing or in pharmaceuticals, and if this is the case, how the future use of biotechnology might conduce to this general end.

## RESEARCH ON BIOTECHNOLOGY AND PARTICIPATION IN TECHNOLOGICAL AND INDUSTRIAL DECISION-MAKING

The reports from this workshop indicated that the participants spent some time debating political ideals. They tended to discuss what scientists or industrial managers or elected parliamentary representatives ought to do, or ought to be able to do. The problem in this context is how to turn such general, debatable statements into research questions. Given that different political actors disagree on major issues like who should control corporate investment, or who should set standards, what kinds of research can move the debate forward and stimulate new lines of argument.

Three general areas suggest themselves, as particularly relevant to biotechnology. Firstly there is the contingent admission of uncertainty or ignorance; secondly the site of origination of concern and thirdly the issue of participants' learning. The first two apply rather more obviously to risk assessment, the last can be extended to democracy in industrial planning and conflict over corporate goals as well.

The debate in the late 1970s over the risks of recombinant DNA research, and its analogue today over the risks to environment of the release of genetically engineered organisms are tremendously instructive as case-studies in the collective management of partial understanding. In the earlier debate molecular biologists began with an admission of ignorance, which they gradually modified, as new information became available. But it is all too clear that it was only by reformulating the issues and shifting the agenda of debate that molecular biologists were able to make the kind of authoritative statements necessary for the production of a consensus that their research was of minimal risk. To them this became increasingly important as the prospect of highly restrictive controls on experimentation moved closer to the statute book. Politically it became increasingly necessary to appear to have minimised the uncertainties, despite the relative paucity of the kind of data initially said to be needed. The adversarial nature of the debate in the

United States forced many scientists to censor their sceptical thoughts, when the public discussion of unresolved questions could have been highly productive.

Interestingly in the less conflictual discussions at present about environmental releases, it is both harder for scientists to claim that the risks are well understood and harder for their critics to claim that major problems are being overlooked. The admission of important residual uncertainty actually seems to be forthcoming from scientists and to have inhibited strong controls on science at this stage. Thus the question of the conditions under which such openness about incomplete information is sustained would be interesting to explore. Linked to this is the question of where historically concern about the risks or the implications of research have been raised. Which social groups have tended to press such issues ?

* * *

This takes us to the final question about learning through participation in conflict and debate. It is just too dispiriting to believe that history is inevitably an endless series of replays, that in each generation or in each country same kinds of conflicts are rehearsed yet again, without any learning taking place, although this is what some superficial and cynical philosophies of history imply.

It may perhaps be so, if it is allowed to be so, but for millenia people have learnt to treat history as a social resource, as a reserve of experience that has to be nurtured and maintained, but which can be drawn upon for guidance, inspiration and enlightenment. This is as true of the history of technology as it is of the history of general enfranchisement. Thus with respect to the recent history of biotechnology we should already ask ourselves about the meaning of the debates and conflicts, to identify those events and interactions that modify attitudes, increase knowledge and shift the arguments forward.

# BIBLIOGRAPHIES

# Bibliography

BIBLIOGRAPHY

Anon, *Scrip,* No 1189, March 20 1987, 26.

Balz, F. et. al., "Production of Novel Antibiotics by Gene Cloning and Protein Engineering" in *Protein Engineering: Applications in Science, Medicine and Industry,* Academic Press, London, 1986.

Barker, S.A., "Early nineteenth century biotechnology", *International Industrial Biotechnology*, vol. 6 October-November 1986, 177-9.

Barna, T., "Process plant contracting: a competitive new European industry" in Shepherd, G., et al., (eds.), *Europe's Industries: Public and private strategies for change,* Frances Pinter, London, 1983, 167-185.

Behbehani A.M.,"The smallpox story: life and death of an old disease", *Microbiological Reviews,* vol. 47 1983, 455-509.

Bennett, P. Glasner, D. Travis, *The Politics of Uncertainty: Regulating recombinant DNA research in Britain,* Routledge and Kegan Paul, 1986.

Beran, G.W., Crowley, A.J.,"Toward worldwide rabies control", *WHO Chronicle,* vol. 37, 1983, 192-196.

Bevan, S., Parsons, D., Pearson, R., *Monitoring the biotechnology labour market*, Science and Engineering Research Council, Swindon, 1987.

Bialy, H., "Biotechnologies converge on new vaccines" *Biotechnology*, vol. 5, January 1985, 11.

Bijman, J., et al., *The Impact of Biotechnology on Living and Working Conditions in Western Europe and the Third World*, English version of report published in Dutch as The *international dimension of biotechnology in agriculture*, European Foundation, Dublin, 1987.

Bodmer, W.F., et al.,"Localization of the gene for familial adenomatous polyposis on chromosome 5", *Nature*, vol. 328, August 13 1987, 614-6.

Botkin, J.W., et al., *No Limits to Learning: Bridging the Human Gap: The Club of Rome Report*, Pergamon Press, Oxford, 1979.

Bud, R., "Continuous brewing and the image of biotechnology in Britain, 1950 to 1970", Paper presented to the conference on *Biotechnology: Long term development*, Imperial College, London, 1984.

California Biotechnology, Annual Report 1986.

Chin, J.,"The status and challenge of communicable disease control" in Last, J.M., (ed.) *Public health and preventive medicine*, Appleton-Century-Crofts, Norwalk, Connecticut, 1986: 103-107.

Colombo, U.,"Nuove technologie, nuovi mestieri", *Notiziario ENEA*, vol. 4, 1984.

Colombo, U., "IDEA: Innovative Dimensions in Energy and Agriculture", *The International Spectator*, vol. 21, no. 4, October-December 1986.

Commission of the European Communities (CEC), *The Agricultural Situation in the Community, 1985 Report*, Office of Official Publications, Brussels/Luxembourg, 1986.

Cossalter, C., *Biotechnologies. Recherche - Emploi - Formation* Centre d'Etudes et des Recherches sur les Qualifications, Paris, May 1986.

Crott, R., *The EEC Policy on Isoglucose*, European Commission, FAST occasional paper No. 41 (XII-1075-81 E) September 1981.

CUBE, *Biotechnology in the Community: Stimulating agro-industrial development*, Brussels, DG XII-CUBE: SDM 3/57, 1986.

Davis, K., "Insurance - an old safeguard for a new technology", *Trends in Biotechnology*, vol. 4, July - August 1984.

Deutscher Bundestag, *Chancen und Risiken der Gentechnologie: Der Bericht der Enquete Kommission,* Bonn, 1987.
Diamond, J.M., "AIDS: Infectious, genetic or both ?" *Nature*, vol. 325, 16 July 1987, 199.

Dimond, P.,"Catholic University's Cell Bio Center Combines Biotech Lectures and hand-on Training", *Genetic Engineering News*, vol. 5, 1985.

Dixon, B., "The effect of the media on public opinion and public policy" in D. Davies, (ed.), *Industrial Biotechnology in Europe: Issues for Public Policy,* Pinter, London, 1986.

Duchene, F., Szczepanik, E., Legg, W., *New Limits on European Agriculture: Politics and the Common Agricultural Policy*, Croon Helm, London, 1985.

Dunnill, P.,"Biotechnology and industry",*Chemistry and Industry,* 4 April 1981, 204-217.

Farrands, C.,"Medicine and health care", *European Trends,* 1984, 22-30.

Flagg, R., Purnell, P., "Hygienic considerations in biotechnology Processes", Paper presented at Online Conference on Biotechnology, May 4-6, 1983.

Robert Fleming Securities, *Pharmaceutical Products Worldwide,* Fleming Research, London, 1987. (1987a)

Robert Fleming Securities, *Overview of viral diseases and viral therapy,* Fleming Research, London, 1987. (1987b)

Foege, W.H., "Banishing measles from the world", *World Health Forum,* vol. 5, 1984, 63-65.

Freeman, C., et al,"Chemical process plant: innovation and the world market", *National Institute Economic Review* vol. 45, 1968.

Fresco, N.,"La modethique" in M. Azoulai, P. Jouannet, (eds.), *L'Ethique: Corps et Ame,* Editions Autrement, Paris, 1986, 168-74.

Fruin, W.M., *Kikkoman: Company, clan and community*, Harvard University Press, Cambridge, 1983.

Fundenberg, H.H., "Fiscal returns of biomedical research", *The Journal of Investigative Dermatology,* vol. 61, 1983, 321-329.

Haber, L.F., "From alkalis to petrochemicals: economic development and technological diffusion in the British chemical industry, 1914-1964" in Okuchi, A., Uchida, H., (eds.), *Development and Diffusion of Technology: Electrical and Chemical Industries*, University of Tokyo Press, Tokyo, 1980, 97-121.

Hacking, A.J., *Economic Aspects of Biotechnology*, Cambridge University Press, Cambridge, 1986.

Hastings, J.J.H., "Development of fermentation industries in Great Britain" in Perlman, D, (ed.), *Advances in Applied Microbiology*, Academic Press, London, 1971, 1-45.

Henderson, D.A., "The eradication of smallpox", in Last, J.M., *Public health and preventive medicine*, Appleton-Century-Crofts, Norwalk, Connecticut, 1986, 129-138.

Hilleman, M., et al.,"Subunit *Herpes simplex* vaccine", in Nahmias, A., Dowdle, W., Schinaze, R., (eds.), *The human herpes viruses: An interdisciplinary perspective*, Elsevier, New York, 1981, 503-506.

Holler, H., Moller, K.M., ( eds.), *The Carlsberg Laboratory 1876-1976*, Carlsberg Foundation, Copenhagen, 1976.

Houwink, E., *A Realistic View of Biotechnology*, DECHEMA for the European Federation of Biotechnology, Frankfurt, 1984.

Institute of Medicine, *New vaccine development, establishing priorities*. Volume 1. *Diseases of importance in the United States*, National Academy Press, Washington, DC, 1985.

Jacobsson, S., Jamison, A., Rothman, H., *The Biotechnological Challenge* Cambridge University Press, Cambridge, 1986.

Jasanoff, S.,"Technological innovation in a corporatist state: the case of biotechnology in the Federal Republic of Germany", *Research Policy*, vol. 14, 1985, 23-38.

Jeffreys, A., et al.,"Positive identification of an immigration test-case using human DNA fingerprints", *Nature*, vol. 317, 31 October 1985, 518-9.

Kalter, R.J., "The New Biotech Agriculture: Unforeseen Economic Consequences", *Issues in Science and Technology*, Autumn 1985, 127.

Kaplinsky, R., "Technological revolution and the international division of labour in manufacturing: a place for the third world", Paper presented to the EADI Conference on New Technologies and the Third World, Institute of Development Studies, University of Sussex, July 1987.

Kenney, M., *Biotechnology; The University-Industry Complex,* Yale University Press, New Haven, 1986.

Klarman, H., Guzick, D., "Economics of influenza", in Selby, P., (ed.), *Influenza: virus, vaccine and strategy,* Academic Press, New York, 1976, 255-270.

Kobbe, B., "Survey examines present and future personnel needs in biotechnology", *Genetic Engineering News,* October 1986.

Koplan, J.P., Kane, M.A., "Assessment of preventive health technology: the case of Hepatitis B vaccine", *Journal of Health Care Technology,* vol. 2, 1986, 217-224.

Kohler, R.E., *From Medical Chemistry to Biochemistry: the making of a medical discipline,* Cambridge University Press, Cambridge, 1982.

Krimsky, S., *Genetic Alchemy: A social history of the recombinant DNA controversy,* MIT Press, Cambridge, 1982.

Lanzavecchia, G., Mazzonis, D., *The impact of biotechnology on living and working conditions,* European Foundation, Shankill, Co. Dublin, April 1986.

Lee, J., *The Impacts of Technology on the Alternative Uses for Land (Final Report),* FAST-Programme, An Foras Taluntas, Johnstown Castle, Wexford, 1985.

Lewis, C., Kristiansen, B., "Chemicals manufacture via biotechnology - prospects for Western Europe", *Chemistry and Industry,* 2 September 1985, 575.

Liebenau, J., "A comparison of penicillin research and development in Britain and the United States", Paper presented to the meeting on Biotechnology: Long Term Development, Imperial College, London, 1984.

McGarity, T. O. "Legal and Regulatory Issues in Biotechnology", *Biotechnology and the Environment: Risk and Regulation Seminar,* 1985.

Meadows, D.H., et al., *The Limits to Growth: A report for the Club of Rome's Project on the Predicament of Mankind* Universe Books, London, 1972.

Miller, H., "Report on the World Health Organization Working Group on Health Implications of Biotechnology", Recombinant DNA Technical Bulletin, June 1983.

Ministry of Economic Affairs, The Netherlands, *Biotechnology in the Netherlands: A strong foundation - a shining future,* Commission on Foreign Investment, The Hague, 1986.

Mulley, A., Silverstein, P., Dienstag, J., "Indications for use of hepatitis B vaccine, based on cost-effectiveness analysis", *New England Journal of Medicine,* vol. 307, 1982, 644-652.

Murray, F., "The decentralisation of production: the decline of the mass-collective worker", *Capital and Class,* vol. 19, Spring 1983, 74-99.

National Science Foundation (NSF), *Industrial Biotechnology R&D Performance,* Washington, 1987.

Office of Technology Assessment (OTA), *A review of selected federal vaccine and immunization policies,* US Government Printing Office, Washington, DC, 1979.

Office of Technology Assessment (OTA), *Compensation for vaccine-related injuries. A technical memorandum,* US Government Printing Office, Washington, DC, 1980.

Office of Technology Assessment (OTA), *Cost-effectiveness analysis of influenza vaccine,* US Government Printing Office, Washington, DC, 1981.

Office of Technology Assessment (OTA), *Commercial biotechnology: an international analysis,* US Government Printing Office, Washington, DC, 1984.

Office of Technology Assessment, (OTA) *Technology, Public Policy and the Changing Structure of American Agriculture* USGPO, Washington, 1986.

Okun, S.,"Personnel availability remains a major concern", *European Chemical News,* Supplement, May 1984.

Otway, H.J., Peltu, M., *Regulating Industrial Risks: Public, experts and media,* Butterworths, London, 1986

Pantell, R.H., Stewart, T.J., "The pneumococcal vaccine: immunization at a crossroad", *Journal of the American Medical Association,* vol. 241, 1979, 2272-2277.

Parsons, D., *Directory of Short Courses in Novel Biotechnology 1986-87,* SERC, Swindon, 1987.

Parsons, D., Pearson, R., *Enabling Manpower for Biotechnology in the UK,* Report for Biotechnology Directorate of the Science and Engineering Research Council: Institute of Manpower Studies, University of Sussex, December 1983.

Pearson, R., "An international biotechnology labour market, *Nature,* vol. 308, 1984, 572.

Pearson, R., Parsons, D., *The Biotechnology Brain Drain,* SERC, Swindon, 1984.

Peckham, B.W. "Technological change in the British and French starch industries", *Technology and Culture,* vol. 27, 1986, 18-39.

Perez, C.,"Structural change and assimilation of new technologies in the social and economic system", *Futures,* vol. 15, October 1983, 357-375.

Perkins, F.T., "Vaccination against communicable diseases", *Symposia Series in Immunological Standardization,* vol. 22, 1973, 1-407.

Rexen, F., Munck, L., *Cereal Crops for Industrial Use in Europe,* Carlsberg Research Laboratory, Copenhagen, 1984 (Report prepared for the Commission of the European Communities, EUR 9617 EN).

Rose, N., *Chaim Weizmann: A biography,* Weidenfeld and Nicolson, London, 1987.

Ruitenberg, E.J., "Vaccins", *in Toekomstverkenningen. Jaarboek Gezondheidsraad 1986,* The Hague, The Netherlands, 59-70.

Ruivenkamp, G.,"The Impact of Biotechnology in International Development: Competition between Sugar and New Sweeteners", in G. Junne (ed.), *New Technologies and Third World Development,* Vierteljahreberichte der Friedrich-Ebert-Stiftung, Special Issue, No 103 (March 1986).

Sargeant, K., "Biotechnology, connectedness and dematerialisation: the strategic challenges to Europe and the Community response", Paper presented to "Biotechnology '84", Royal Irish Academy, Dublin, April 1984.

Sargeant, K. (CUBE), Evidence before the Select Committee on the European Communities, House of Lords, Session 1986-87, on 25 Feb 1987 in *Biotechnology in the European Community,* Select Committee Report, HMSO, London, 1987.

Scheer, P.L., "The potential impact of biotechnology on poultry breeding", *Veeteelt,* November 1985, 989.

Schogt, J.C.M., Beek, W.J., (eds.), *The future of our food industry,* Elsevier, Amsterdam, 1985.

Science and Engineering Research Council (SERC), *Biotechnology in Japan,* SERC, Swindon, 1987.

Shaffer, P., "The architect's strategic role in planning and designing facilities for biotechnology", *Genetic Engineering News,* March - April 1983.

Sigerist, H., *Man and medicine,* McGrath Publishing Company, College Park, Maryland, 1970.

Sigsworth, E.M.,"Science and the brewing industry, 1830-1900", *Economic History Review,* vol. 17, 1965, 536-550.

SRI, *Biotechnology in Agriculture. Advances in commercial livestock and plant production technology*, SRI International (Business Intelligence Program), Report No. 7 (Fall 1984), 6-12

Strickland, S.P., *Politics, Science and Dread Disease: A short history of the United States medical research policy*, Harvard University Press, Cambridge, 1972.

Teich, M.,"Fermentation theory and practice: The beginnings of pure yeast cultivation and English brewing", *History of Technology*, vol. 8, 1983, 117-33.

Van Brunt, J.,"Contract production: buying technical expertise", *Biotechnology*, vol. 4, August 1986, 701-5.

Van Kasteren, J., "A drop of soya in your coffee", *NRC Handelsblad*, 19 June 1985, 19.

Webber, D., "Biotechnology firms reshape staffs for move into marketplace", *Chemical and Engineering News*, 18 June 1984.

Weir, R.B. "Distilling and agriculture, 1870-1939", *Agricultural History Review*, vol. 32, 1984, 49-62.

Weisbrod, B.A., Huston, J.H., *Benefits and costs of human vaccines in developed countries: an evaluative survey*, Madison, Wisconsin: Center for Health Economic and Law, University of Wisconsin-Madison, Discussion Paper No. 210, December 1983.

Willems, J, et al., "Cost-effectiveness of vaccination against pneumococcal pneumonia", *New England Journal of Medicine*, vol. 303, 1980, 553-559.

Wolf, E.C.,"Conserving biological diversity" in Lester R. Brown and colleagues, *State of the World 1985: A Worldwatch Institute Report on Progress toward a Sustainable Society*, Norton, New York, 1985.

Wyke, A.,"Molecules and markets: a survey of pharmaceuticals", *The Economist*, February 7 1987, Supplement, 1-18.

Yoxen, E.J., *Unnatural Selection ? Coming to terms with the new genetics*, Heinemann, London, 1986.

Yoxen, E.J., *The impact of biotechnology on living and working conditions*, European Foundation, Shankill, Co. Dublin, 1987.